*Engineering Applications –*
*A Project-based Approach*

# Engineering Applications – A Project-based Approach

**James Ritchie**

**Graham Simpson**

*Department of Mechanical and Chemical Engineering, Heriot-Watt University, Edinburgh, UK*



BUTTERWORTH
HEINEMANN

Butterworth-Heinemann
Linacre House, Jordan Hill, Oxford OX2 8DP
225 Wildwood Avenue, Woburn, MA 01801-2041
A division of Reed Educational and Professional Publishing Ltd

A member of the Reed Elsevier plc group

OXFORD BOSTON JOHANNESBURG
MELBOURNE NEW DELHI SINGAPORE

First published 1998



**British Library Cataloguing in Publication Data**
Ritchie, James
Engineering applications: a project based approach
1. Engineering
I. Title II. Simpson, Graham
620

ISBN 0 7506 2577 5

Composition by Genesis Typesetting, Rochester, Kent
Printed and bound in Great Britain

# Contents

# Preface

Engineers are at the core of civilisation as we know it. Without engineers the world would not be the place it is today, in both the context of the good and the bad in society and all the man-made things we see around us. The responsibility that engineers carry as they design and make products, develop systems and do analyses is immense. People's lives and livelihoods depend on engineering decisions; no-one can escape the influence of engineering in today's modern world. Therefore, to hold the position of a practising engineer can be both daunting and exciting.

The 1980s and 1990s have seen the advent of aggressive world-wide industrial competition as companies adapt to meet the challenge of international competition for their markets, both at home and abroad. At the centre of this revolution is the way in which engineered products are created, designed, marketed, sold and manufactured by engineering companies. Engineers must have an appreciation of how products can be designed for manufacture, how they need in-built reliability and quality, how they must meet customer needs and how these requirements can be met. The amazing rise of Japan as an economic superpower in the last thirty years has been mainly due to the way in which their engineers understand and practise all of these disciplines and are involved at all levels of any manufacturing company. There is an increasing emphasis on the design and manufacturing functions within modern industrial companies to help them compete and achieve world-class status. In the modern world of business the engineer may not be the king but is at least the king's right-hand man.

Engineering design, problem-solving and manufacturing are the areas which every engineer must appreciate if he or she is to contribute in an effective manner to their discipline, regardless of what type of engineering they practise. Not only the technological problems must be addressed but also those influencing cost, people, the environment and society. Other skills in the areas of business, commerce, computing, systems analysis, personnel, etc., must also be developed.

This book is intended for use by students who are pursuing a course which involves engineering design and manufacture. Its main emphasis is on design for manufacture through a series of projects which range from a simple, single item to more complex assemblies. Each of these design briefs will take the student through the basic process of engineering a product, i.e. design, plan for manufacture, manufacture and test. The intention is to develop an appreciation of the importance of designing products for manufacture so that they can be made effectively using the resources available.

Each project includes most of the associated engineering activities such as drawing office work, manufacturing technology, engineering materials selection and treatment and production planning, culminating in the manufacture and test of the final product. It is hoped that the projects will give an introduction to these processes as well as providing a good foundation for more challenging design project work in the future.

Some of the projects are not design-related, e.g. the assembly strip-down exercise and the financial analysis. However, these are deliberately included to provide experience of other skills and knowledge required by the modern engineer – presentational and analytical skills, financial skills, etc.

## So, why have we chosen the title *Engineering Applications*?

In the UK, Engineering Applications should be central to the engineering content of any accredited Chartered Engineering course. It is divided into two sections, namely:

(a) EA1: an introduction to the fabrication and use of materials
(b) EA2: applications of engineering principles to the solution of practical problems.

The projects in this book cover material in both areas and are suitable for incorporation in any fully accredited engineering degree course.

The book contains five main chapters, and a Quick Reference Guide. Each of these will now be briefly explained in turn.

**Chapter 1** defines products in a general sense in the context of the modern manufacturing market and the environment in which companies who make products now find themselves competing. It also outlines the cycle of 'Engineering a Product', emphasizing the importance of design for manufacture (DFM) and the benefits it can give a manufacturing company if applied properly. Further, it explains the manufacturing system and typical departments associated with product development and manufacture in any company along with some of the activities carried out and the documentation produced.

**Chapter 2** outlines the process of design and drafting and how it has been managed in a traditional manufacturing environment. It then explains how modern methods, which are being increasingly adopted, are helping some companies thrive in the international marketplace. A general methodology for approaching the projects is outlined along with instructions on how the design information must be presented.

**Chapter 3** briefly outlines different forms of manufacturing and the need for carefully prepared manufacturing information before product production takes place. Advice is also given on useful background information material which will help students to tackle the manufacturing parts of their projects.

**Chapter 4** outlines the function of process planning. One of the key documents produced for any product is the process plan; this outlines the sequence of operations to transform the item from its raw material state to the finished component or assembly. It also formalizes a methodology for creating process plans and presents a few examples to help in the development of skills in this area.

**Chapter 5** outlines the projects. These are given numbers from Type 1 through to Type 10, the degree of difficulty tending to get higher as they are completed. The outlines are designed to support the project specifications provided in the *Tutor Resource Pack*, which should be given to you as handouts as you tackle each project type.

The **Quick Reference Guide** (Part Two) contains not only material for use during engineering applications projects but also useful data and information which may be required in the rest of the course and during a career as a professional engineer.

Finally, we hope that students will enjoy this first experience of designing and manufacturing in the world of engineering and we hope that it will confirm their wish to become a professional engineer. We are sure that what is learnt by carrying out these projects will be relevant and useful in the future – good luck!

# Acknowledgements

Extracts from British and International Standards are reproduced with the permisson of BSI. Complete copies can be obtained by post from BSI Customer Services, 389 Chiswick High Road, London W4 4AL (telephone 0181-996 7000) or through national standards bodies.

Figure 1.4 (page 7) is an adaptation of Figure 4.3 (page 57) from *Production and Operations Management Text and Cases*, Terry Hill. Reproduced with the kind permission of Prentice-Hall Europe, 6 Wood Lane End, Hemel Hempstead, Hertfordshire HP2 4RG.

The expanded planning method used in pages 38–40 is an adaptation of that given in the book *Introduction to Jig and Tool Design*, M.H.A. Kempster. Reproduced with the kind permission of Edward Arnold (Publishers) Ltd, 338 Euston Road, London NW1 3BH.

The authors also wish to acknowledge the advice and assistance obtained from their many colleagues at Heriot-Watt University who have been involved in developing mechanical engineering EA teaching programmes. In particular they would like to thank Ian Black, Norman Loch, Brent Wilson, George Hartnup and Bruce Davies for their input, support and encouragement over the years.

# Part One

## Engineering Applications

# 1 Engineering a product

The engineer in today's society must be aware of how products are marketed, researched, designed, developed and manufactured. The engineering of products involves a wide range of activities, all of which have to interact efficiently and effectively such that companies can produce products which compete in the modern marketplace. As a consequence of this, engineers must be aware of all stages of product development and manufacture as well as the relevant business aspects associated with these. They also need to have a broad understanding of how a modern company operates and competes as well as having detailed knowledge of specialist areas. Without the former it will be impossible to apply the latter successfully and to its full advantage.

## 1.1 Products

Engineers need to develop and manufacture products, but what are they?

Products are consumable goods, items which are bought and used by people depending on what they desire and/or need; they may buy the goods for their own use or on behalf of an organization or company. All manufactured goods have a limited life-span, some of a few seconds, others of many years. Ultimately, one way or another, their useful life will come to an end.

By simply looking around your surroundings you will see examples of products – furniture, pens, clothes, dishes, newspapers, to name but a few. All of these have been manufactured to be sold at a specific price which has to ensure a profit for the company which manufactured them and the retailer who sold them, after they have paid back their investment costs.

Some products, such as cutlery and screws, are very simple in form, consisting of only one part. Others can be either simple or complex assemblies ranging from electrical plugs, light switches and door handles to motor cars, aeroplanes and television sets.

These may be obvious examples of end products but it should be appreciated that the more complex products may themselves be combinations of simpler products. Just as bricks are a product bought to be combined together into a building which will in turn be purchased as a finished product so, for example, electronic components are the building blocks purchased by companies manufacturing electronic products for sale in the marketplace.

A manufacturing company may not make all, or indeed any, of the detail parts which go into its final assembled product. There is nearly always a need for them to buy parts or raw materials from another company, usually called a supplier or a vendor.

All products require the input of professional engineers at many stages, not only in the design and manufacture of the product itself but also in the development and manufacture of the machines and other ancillary equipment used to make it. Examples are machines for cutting and sewing cloth, machine tools for producing metal components for cars, and computers used for controlling manufacturing processes and ordering parts.

As mentioned previously, all products are bought and sold in the marketplace. When you see an item of clothing, a compact disc, a computer game or a video which you then decide to buy, you are, in your own small way, stimulating a demand for that product. If many other people also decide to buy it and it becomes more popular than any other competing product then it becomes a market leader. This gives the manufacturer or retailer an excellent chance of maximizing its profits before the product's life cycle comes to an end. Demand for a product may fall for a number of reasons, such as it going out of fashion, anyone who was going to buy it has bought it, or competing products enter the market at a better price and quality.

A product life cycle is considered to be the time from when it is introduced to the market until it is withdrawn. As shown in Figure 1.1 the traditional product life cycle was usually very long. This enabled a business to make profits over a substantial period of time.

The modern product life cycle is often quite different. Due to increasing and aggressive worldwide competition many manufactured products now have a shorter life cycle, and therefore it is necessary for

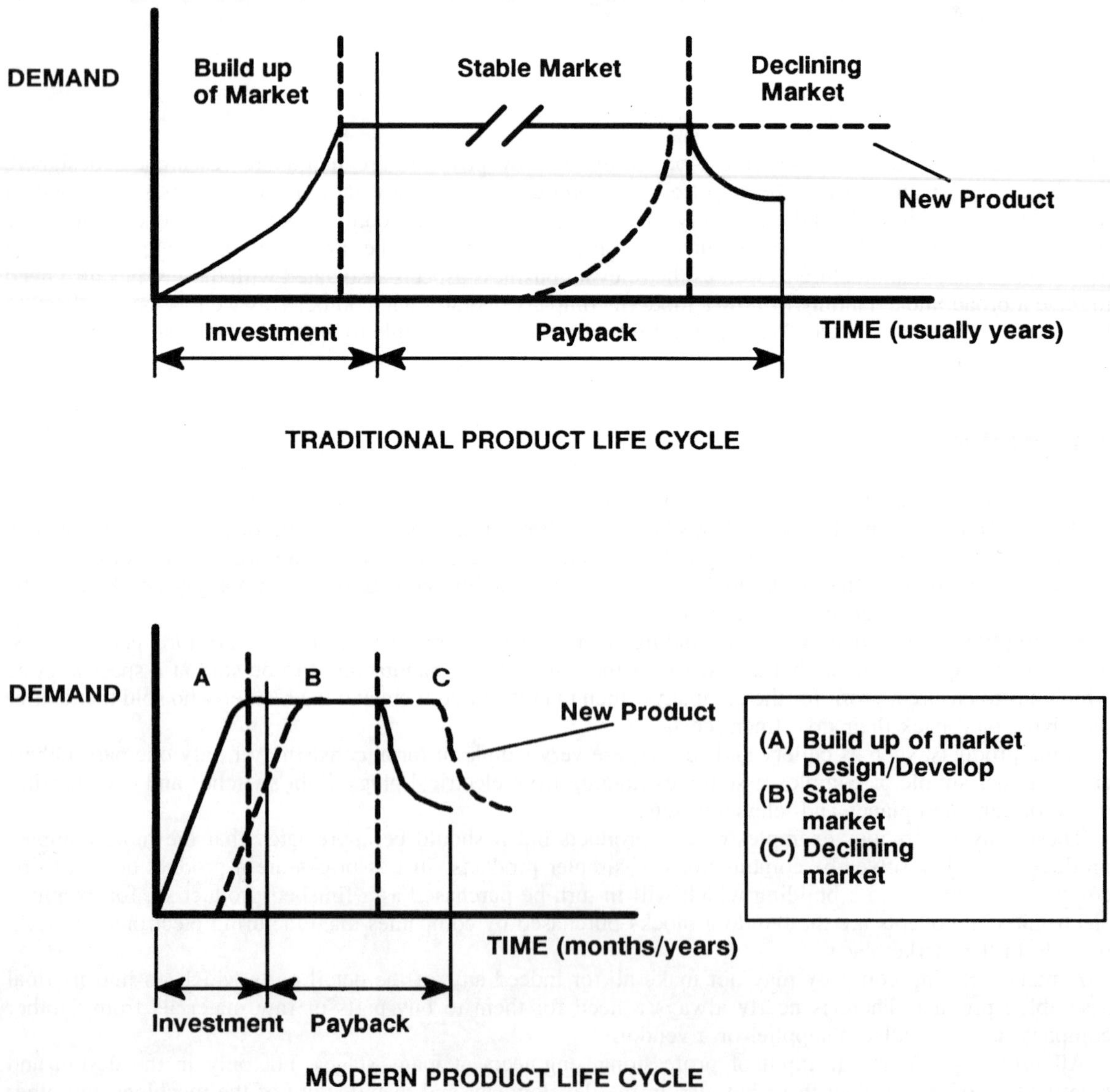


**Figure 1.1** *Idealized product life cycles*

companies to design, develop and manufacture their product quickly and with good quality so that they can beat the competition to the marketplace and, most importantly, satisfy the demands of the customer. Then new products or variants of existing products must continuously be developed (quickly) and introduced so that a market lead is maintained. The engineer has a key role to play in this process.

The vast majority of products engineered and developed in the modern world need a market. No longer can engineers develop products just because there may be an engineering or technical challenge; the market must be investigated first. If it is found that a demand for a new product exists, then, and only then, is the product developed in detail. Engineers must recognize that it is important to obtain the advice of experts from sales and marketing before the detailing stage.

Thus engineers must be capable of applying their skills and knowledge to the effective development of new or existing products so that the company they work for can survive in the modern marketplace and keep or achieve a market lead. New designs need to be manufactured at a cost-effective price and quality. It is therefore important that new products are designed to suit the manufacturing processes available as far as possible. The concept of design for manufacture (DFM) is a key awareness which it is hoped EA projects will give to the modern young engineer. This awareness should not be viewed in isolation but in the context of a company trying to compete in the modern manufacturing market; the advantages DFM can give it are then apparent.

No matter how good the design of a product, if it cannot be manufactured easily, cost effectively and with good quality, then the competition will soon catch up and deliver a better, more marketable product before the required market share is obtained.

## 1.2 The process of engineering a product

The traditional or sequential approach to engineering products is shown in Figure 1.2. Although it is implied from this diagram that all stages of product development should be carried out in this sequence, many modern companies actually carry out some of these functions in parallel. They do this by using the philosophy of simultaneous or concurrent engineering [1,2]. This applies the team approach to the creation of new products, bringing together 'experts' from design, manufacture, quality, marketing, etc. to work together, thus enabling rapid product development. However, for the purpose of developing a detailed understanding of the activities which take place during the engineering of a product, this text will look at each stage individually and sequentially. Once EA projects are tackled in this way it is hoped that the student will see the relationships between each part of the diagram and consider two or more stages together, thus becoming aware of the need for and advantages of DFM and concurrent engineering.

After Marketing and Sales have identified the demand for a product, the engineering departments then carry out the engineering process such that a competitive, cost-effective solution is obtained. These loops may be gone round several times, either wholly or partially, depending on the product's engineering

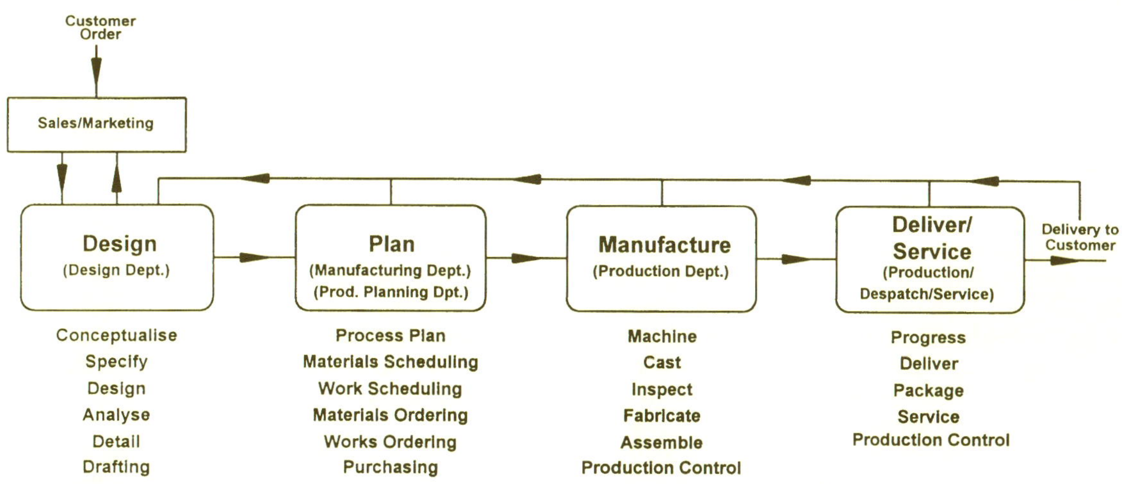


**Figure 1.2** *The process of engineering a product*

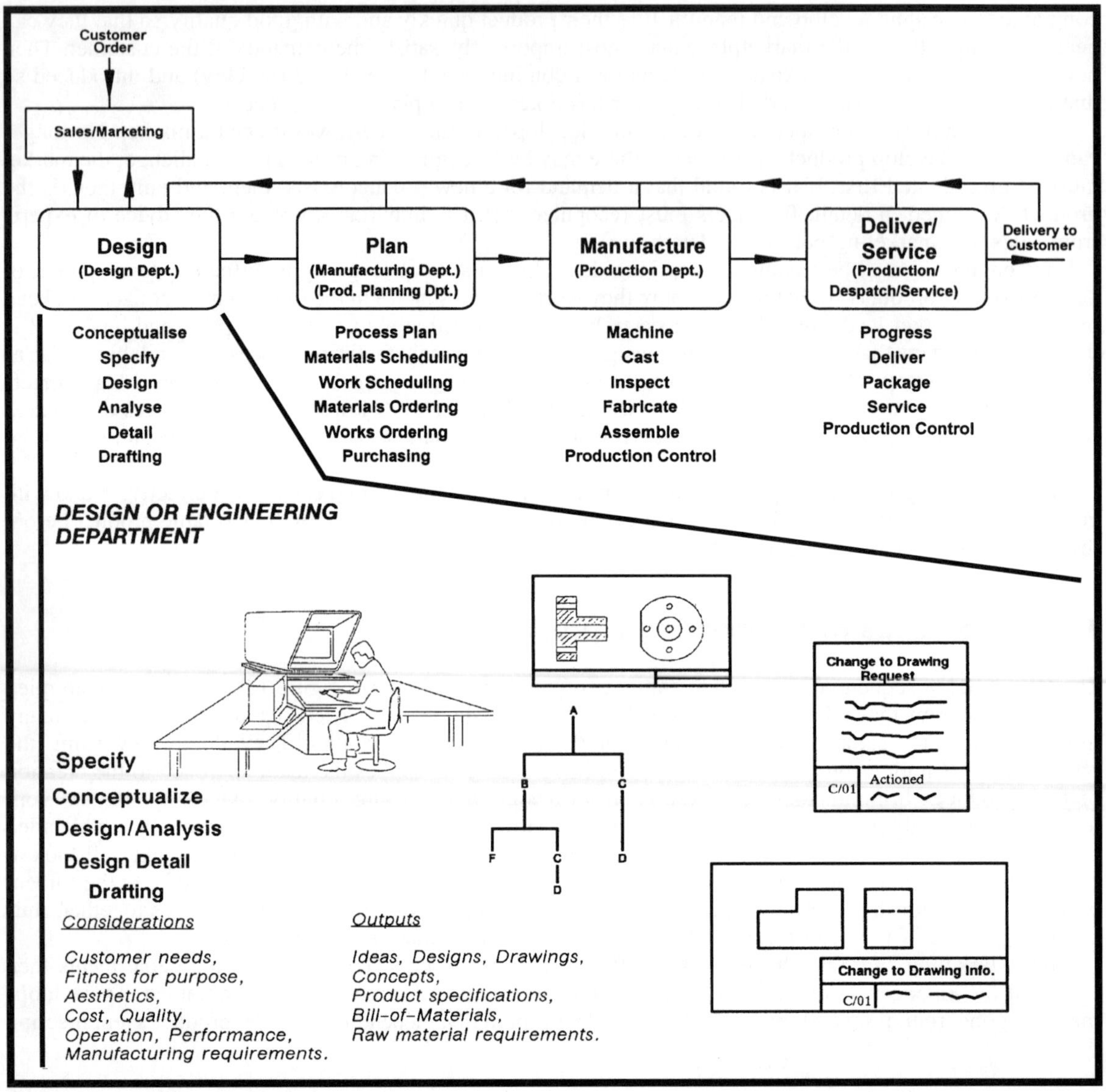


**Figure 1.3** *The design function*

requirements. The feedback loops in the diagram represent recommendations for product improvement or modification due to some need unforeseen at an earlier stage during the engineering process, e.g. a modification to the design requested at the 'plan' stage to make it easier to produce.

The first stage of the 'Engineering a Product' process is called *Design* (Figure 1.3).

At the marketing stage the Design Department may have produced preliminary drawings or sketches. However, once the company has decided to proceed with a new product, this department will specify, develop, analyse and create it with particular regard to its operational, aesthetic and manufacturing requirements. At the end of this stage, final drawings and bills-of-material are available which are then used by the downstream manufacturing functions for product planning and manufacture.

It is important to note that as much as 80 per cent of a product's final cost can be committed at the design stage. This is crucial to the eventual competitiveness of a product in the marketplace and will be discussed in more detail in Chapter 2.

When the design of a product is completed, the manufacturing process begins: this involves both planning manufacture and manufacture itself. The manufacturing process can be examined by investigating the concept of the manufacturing system, which may be considered as a transformation process (Figure 1.4). This can be

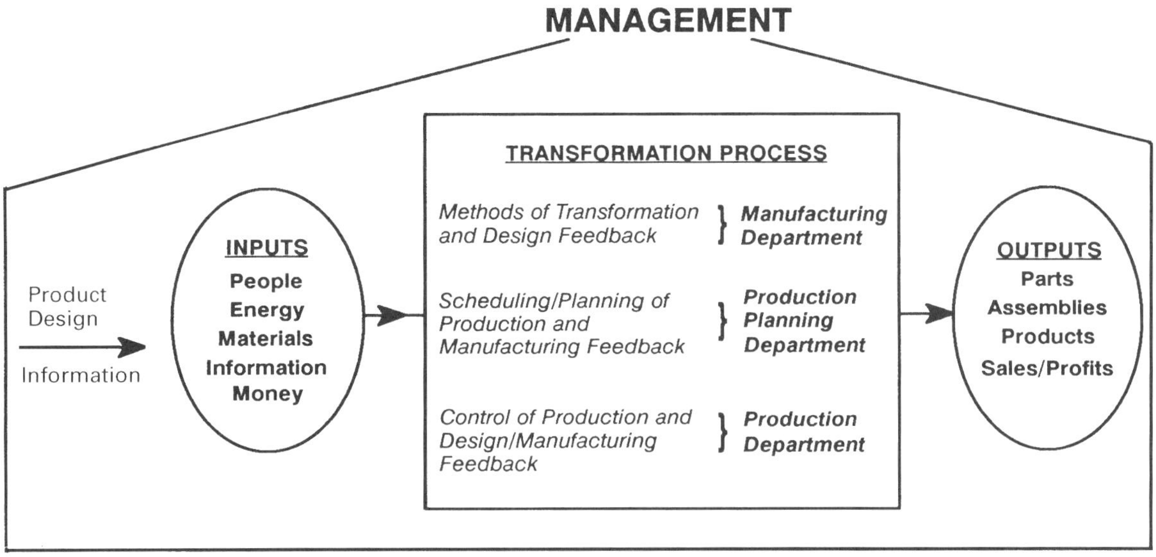


**Figure 1.4** *The manufacturing system*

viewed as any other physical engineering system. On receipt of the design information it involves the inputs of people, energy, materials, money and information being transformed into finished products. These may take the form of individual parts or assemblies. Management has the all-embracing responsibility for organizing the inputs and the transformation process such that the desired outputs are achieved.

There are usually three separate departments involved in the manufacturing system process, namely the Production Planning Department, the Manufacturing Department and the Production Department. In certain companies these titles and their roles may be swapped or combined; however, for the purpose of this book and EA projects these are the definitions which shall be used. The *planning* stage is carried out in two areas by two different departments, namely the Manufacturing Department and the Production Planning Department (Figure 1.5). The Manufacturing Department, which may also be known as the Process Planning Department, prepares and plans detailed manufacturing information for each detailed part or assembly. The personnel involved in this department may be called Production Engineers, Process Planners, Part Programmers, etc. However, their fundamental task is the same, that of converting designs into manufacturing process plans for use on the factory floor by experienced operators.

One of the key tasks in the EA projects outlined in this book is to prepare process plans for selected parts designed by the student.

On receiving a finished drawing and bill-of-material, from Design, the process planners in the Manufacturing Department would determine:

(a) if the item is to be made in-house or purchased from a supplier
(b) what manufacturing facilities and processes are required to produce the item
(c) the sequence of operations required along with the manufacturing resources and tooling needed at each stage – some 'design' of tooling and fixturing may take place at this stage
(d) raw material requirements, sequence manufacturing times and cost estimates
(e) any recommended design changes to facilitate manufacture and liaison with the Design Department.

The chosen manufacturing method would depend on:

(a) the people, machines, processes, tooling and manufacturing system available
(b) the raw materials recommended and used by the company
(c) the drawing geometry, tolerancing and finish requirements
(d) the estimated manufacturing time and cost

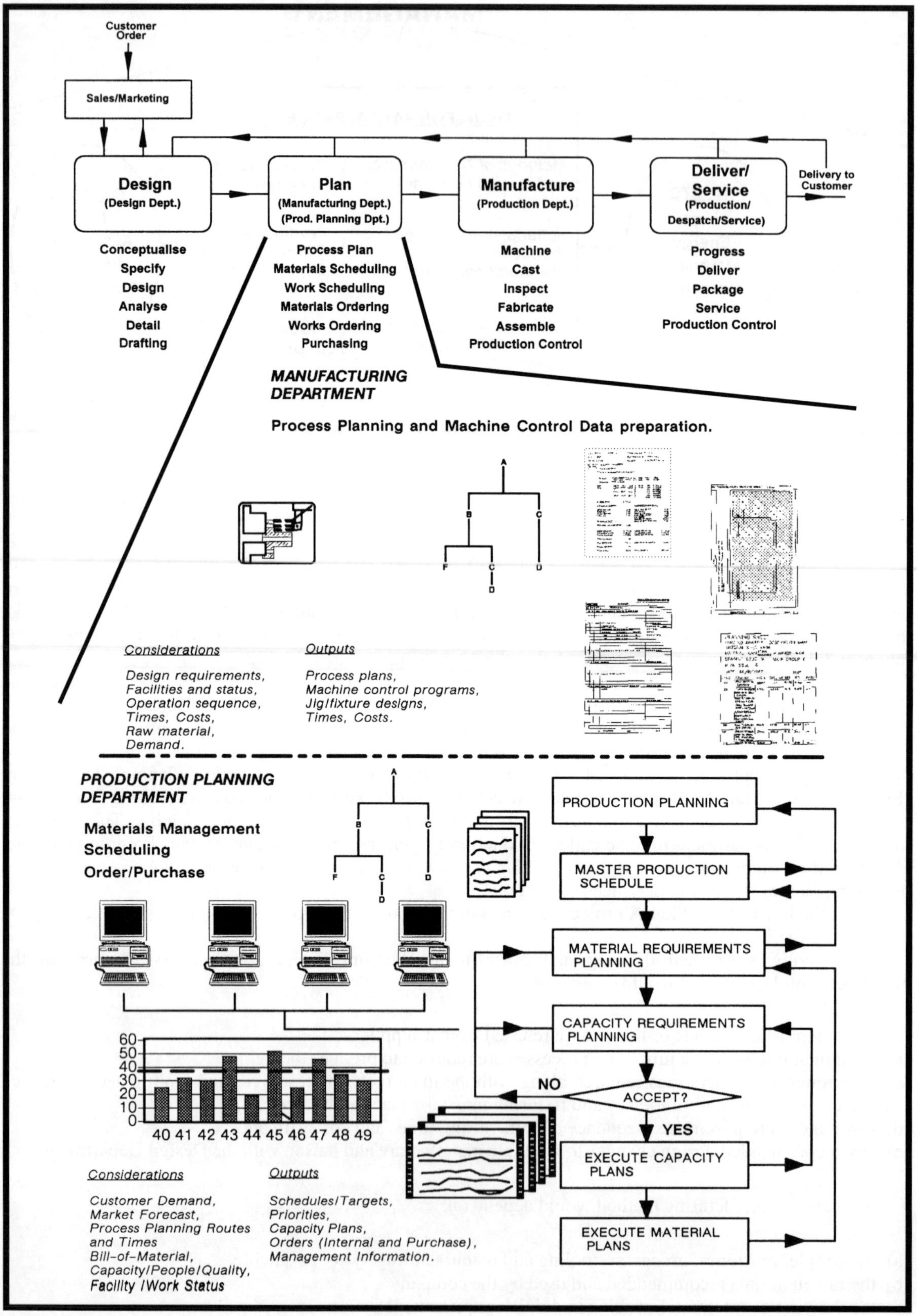


**Figure 1.5** *The 'plan' for manufacture function*

**PARS ENGINEERING LTD.**

**WORKS ORDER**

| | | | |
|---|---|---|---|
| Works Order No. – | **WO1022** | Due Date – | **15 May** |
| Part/Drawing No. – | **PA 1450-000** | Quantity Ordered – | **105** |
| Description – | **PUMP ROTOR** | Quantity Outstanding – | **95** |

**PARS ENGINEERING LTD.**

**East End Industrial Estate**
**Dunfermline**
**Fife KY12 3ED**
**Tel. (0383)654234**
**Fax. (0383) 654896**

**PURCHASE ORDER**

Purchase Order No. – **PO1022**

Date Ordered – **15 February**

To: **Spare Parts Ltd.**
**Donibristle Industrial Estate**
**Dunfermline**
**Fife KY11 2ER**

| Item | Part/Drawing No. | Description | Unit of Quantity | Quantity/ Unit Price(£) | Price (£) | Due Date |
|---|---|---|---|---|---|---|
| 1 | HA 1000-000 | Pump control unit PLC | Each | 150/£20.00 | 375 | 15 May |
| 2 | BJ 2340-123 | Seal Rings | Each | 200/£0.50 | 100 | 15 May |
| 3 | FL 3423-009 | Oil Lubricant – Type 3243 | Litres | 500/£1.20 | 600 | 15 May |
| 4 | DH 5723-900 | 15mm Diameter Copper Cable | Metres | 500/£1.10 | 550 | 15 May |

The above items should be marked for the attention of **Mr R Paton**.

**Total Amount £1625**

Delivery should be made to **GATE 4**.

**Figure 1.6** *A typical works order and purchase order*

(e) the quantity or average batch size of the items being produced for stock or demanded by the customer
(f) the finished order delivery schedule
(g) the status of the production facility.

The main output from this activity is a document known as the *process plan* (see Chapter 5). This specifies the sequence of operations to be carried out in order to manufacture the constituent parts of the product from the procurement of raw material to its completion. For computer-controlled machine tools and processes, e.g. a computer numerical control (CNC) lathe or CNC robot, a machine control program (or part program as it is usually called) may also be produced at this stage and referred to on the process plan.

The Production Planning Department plan, which may also be known as the Materials Planning Department or Production Scheduling Department, schedule and order all company products and constituent parts to meet the targets set by Management. Final product delivery schedule dates and quantities would be set by company management depending on current orders and market forecasts. Using these schedules the Production Planning Department would order all product parts – whether manufactured in-house or purchased from suppliers – and raw materials such as ferrous and non-ferrous metals, plastics, paints, etc. Each order would have a target quantity and delivery date.

For in-house orders production schedules will then be developed giving order- or batch-quantities and target dates along with machine, process or assembly priority work-to lists for the shop-floor operators. Times determined during the process planning stage would be used to help develop these. All schedules and documents may be produced in a modern company using a computer-aided production management (CAPM) system, also known as a manufacturing resource planning (MRPII) system.

Typical company internal and external orders are shown in Figure 1.6. Internal orders may also be known as works orders and external orders as purchase orders.

**** TRAVELLER ****

**PARS ENGINEERING LTD.** **WORKS ORDER TRAVELLER**

Works Order No. - **WO1022** Due Date - **15 May**

Part/Drawing No. - **PA 1450-000** Batch Quantity - **105**

Description - **PUMP ROTOR**

Raw Material: - **50mm Stainless Steel x 0.8m length = 1 component**

Company Inspection standard - **INT23.12**

| Operation Number | Work Centre | Operation Description | Set-up Time (hours) | Run Time (hours) |
|---|---|---|---|---|
| 10 | 2010 | Machine rotor shaft to finished drawing dimensions. | 0.5 | 1.2 |
| 20 | 4310 | Machine rotor shaft keyway to finished drawing dimensions. | 0.25 | 0,45 |
| 30 | 2361 | Finish grind shaft to drawing dimensions and finish. | 1.5 | 0.55 |
| 30 | 2361 | Final inspect to drawing | 1.0 | 0.55 |

**Figure 1.7** *Typical traveller (or route card/batch record card)*

**PARS ENGINEERING LTD.** **** JOB CARD **** **WORKS ORDER JOB CARD**

Works Order No. – **WO1022** Due Date – **15 May**

Part/Drawing No. – **PA 1450-000** Batch Quantity – **105**

Description – **PUMP ROTOR**

| Operation Number | Work Centre | Work Centre Description | Set-up Time (hours) | Run Time (hours) |
|---|---|---|---|---|
| 10 | 2010 | Centre Lathe | 0.5 | 1.2 |

**Operation Description**

Machine rotor shaft to finished drawing dimensions.

**** SET-UP TIME ****

| Date | On | Off | Date | On | Off | Date | On | Off |
|---|---|---|---|---|---|---|---|---|
| | | | | | | | | |

**Setter:** **Setter:** **Setter:**

**** RUN TIME ****

| Date | On | Off | Date | On | Off | Date | On | Off |
|---|---|---|---|---|---|---|---|---|
| | | | | | | | | |

**Operator:** **Operator:** **Operator:**

**PARS ENGINEERING LTD.** **** JOB CARD **** **WORKS ORDER JOB CARD**

Works Order No. – **WO1022** Due Date – **15 May**

Part/Drawing No. – **PA 1450-000** Batch Quantity – **105**

Description – **PUMP ROTOR**

| Operation Number | Work Centre | Work Centre Description | Set-up Time (hours) | Run Time (hours) |
|---|---|---|---|---|
| 20 | 4310 | Vertical Milling Machine | 0.25 | 0,45 |

**Operation Description**

Machine rotor shaft keyway to finished drawing dimensions.

**** SET-UP TIME ****

| Date | On | Off | Date | On | Off | Date | On | Off |
|---|---|---|---|---|---|---|---|---|
| | | | | | | | | |

**Setter:** **Setter:** **Setter:**

**** RUN TIME ****

| Date | On | Off | Date | On | Off | Date | On | Off |
|---|---|---|---|---|---|---|---|---|
| | | | | | | | | |

**Operator:** **Operator:** **Operator:**

**Figure 1.8** *Typical job cards*

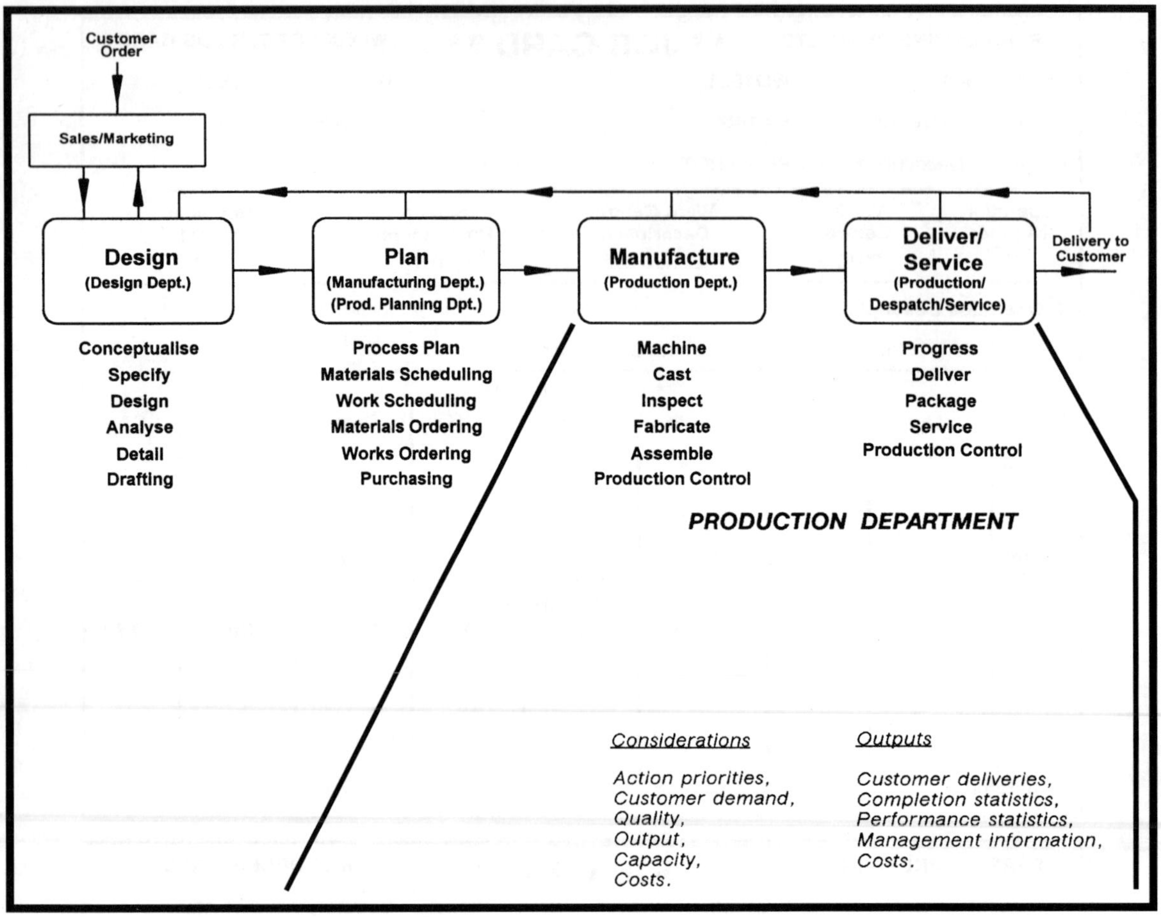


**Figure 1.9** *The 'manufacture/deliver' function*

On releasing works orders to the shop floor, a traveller (Figure 1.7 – this may also be known as a route card or a batch record card) and/or individual operation job cards (Figure 1.8) may be issued together with a material requisition for stores, a copy of the process plan and the appropriate engineering drawings. The traveller will follow the batch throughout its manufacture and the job cards will be completed by the operator as each operation in finished. Modern CAPM systems sometimes allow bar-code reading of finished operation information. Batch manufacture can then begin.

The final two stages in the 'Engineering a Product' process are *Manufacture* and *Delivery*. These are carried out by the Production Department (Figure 1.9). The Production Department, on receipt of the batch paperwork, manages the transformation of raw materials and/or detail parts into products. They arrange to obtain the raw materials or assembly parts, then begin the actual manufacture of the product. This manufacture takes place in a factory which consists of all the machines, processes, production control systems, stores, offices, equipment and people needed to produce the final range of products.

The Production Department's main tasks include:

(a) organizing tooling and raw materials
(b) running and managing the manufacturing facility, its people and plant
(c) meeting scheduled work targets organized by the Production Planning Department
(d) controlling and monitoring work-in-progress (WIP), fixing day-to-day priorities
(e) reporting on progress to the Production Planning Department
(f) feeding back recommended design and manufacture changes to the Design and Manufacturing Departments
(g) reporting on progress or problems to management.

In any company which sells products made in-house, manufacturing employs a large number of people relative to other functions, such as sales, marketing, administration. Many other costs arise due to investment in inventories, e.g. raw materials, tooling, WIP, finished parts storage [3]. Therefore the cost of each of these should be kept to a minimum consistent with a smooth flow of production. Many of these costs can be designed out at the design stage.

The final stage of engineering a product is delivery and servicing. Obviously, many products are delivered and not serviced but where installation, servicing and product support are involved, e.g. with boilers, radar, aircraft, motor cars or central heating systems, a great deal of information is available to a company from service technicians and customers with regard to how the product is functioning and performing. Suggested improvements from users and support staff can greatly enhance the engineering of the overall product and make it more marketable. This is one area which engineering companies do not exploit enough in the 'engineering a product' cycle. Product performance feedback is crucial in determining the market taste, the direction of new products and the improvement of existing products.

Management oversee the running of the plant and all the associated departments. They set targets for production and allocate the resources in terms of money, people, buildings, machines, process requirements, offices and other resources (see Figure 1.4). Typical management structures in companies are not discussed in this text but this is something you should be aware of, particularly for any EA projects relating to investigating companies or engineered products.

Most importantly, the key output in any manufacturing company is sales, without which it cannot pay back any investment, produce profits or make money, which is the ultimate goal of any manufacturing business [3].

Students working on EA projects find themselves in a unique engineering environment. Instead of a number of different departments and people being involved in the design, development, planning and production of products, as is the case throughout a typical manufacturing company, the student alone or as part of a small group is responsible for all stages of design and manufacture. You will design and draw each individual component and assembly, create process plans for their subsequent manufacture and then actually make some or all of the items to produce the final product.

As you go through each stage of product development it is hoped that you will gain an appreciation of each of the functions carried out within a larger organization, particularly in terms of information creation and the responsibility this places on the modern engineer to produce quality information which can be used to enhance the downstream manufacturing functions and hence product competitiveness. Do not forget to include the role of the customer and ask yourself, 'am I happy with this product?'.

## References


1 *Concurrent Engineering – The Product Development Environment for the 1990s*, Carter and Baker, Addison-Wesley (1991)

2 *The Mechanical Design Process*, David G. Ullman, McGraw-Hill (1997, 2nd edition)

3 *The Goal*, E. M. Goldratt and J. Cox, North River Press Inc. (1993, 2nd edition)

# 2
# Design and drafting

This chapter gives an outline of the design and drafting processes as well as the information ready for use after they have been carried out. It is not intended here to teach you design and drafting as subjects – many useful texts already exist for that purpose [1,2,3] – but to make you aware of their key role within the engineering of products. As you attempt the EA projects in this book, much of the intuitive knowledge associated with design should become apparent. It is important to note that up to 80 per cent of a final product's cost is committed at the design stage (Figure 2.1). When an engineering project is financed, very little is spent in the early phase but the decisions taken at this stage determine the eventual product cost. So although very little money is actually *spent* at this point in time, a very large proportion of it is *committed*. The importance of the conceptual design phase of any project cannot be overestimated.

As mentioned previously, prior to the design stage, Marketing will have determined the demand for the product on the basis of a rough general outline; this may include a few sketches. When a policy decision is taken by a company to develop, manufacture and sell a product then it is at the design stage that a detailed

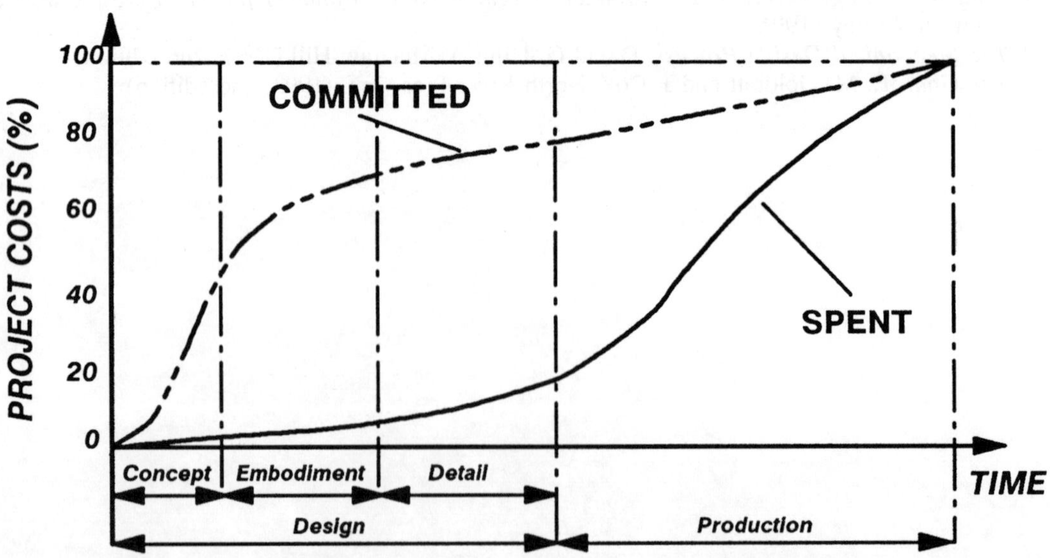


**Figure 2.1** *A graph comparing committed costs and spent costs over the lifetime of engineering a typical product (reproduced courtesy of Ian Black, Heriot-Watt University)*

specification is produced and the initial detailed product information created. This will be used by many of the downstream manufacturing functions for tasks such as process planning and production planning. The more accurate the designer can make this information, the more effectively these areas will be able to complete their tasks, therefore their responsibility is very great.

Considerable reductions in manufacturing costs can be achieved if products are designed to suit the capabilities of the manufacturing systems and processes used by a company. Therefore it is crucial that Design Engineers communicate and discuss their ideas with Manufacturing Engineers so that the full benefits of reusing well created design information can be obtained. The EA projects in this book are developed to encourage you to take responsibility for creating information during all stages of engineering a product. From this you should develop an awareness of this need for communication and collaboration during product development.

The creation of a design within a company usually takes place in a drawing office or DO (Figure 2.2). On entering such a place, which is usually a large room, many desks, drawing boards, people, paperwork and possibly computer-aided design (CAD) terminals will be apparent. The department itself may be titled 'Design', 'Design Engineering', 'Engineering' or similar.

**Figure 2.2** *A typical drawing office (acknowledgements to Cochran Boilers, Annan)*

Design Engineers specify, design, analyse and detail product information. Draftsmen then produce the finished drawing according to the formal company and/or national standard. These drawings are in fact the detailed specification of component parts and assemblies and, once released, the information they convey is then reused by downstream processes to create manufacturing information and instructions.

As you create your designs and associated drawings you will, at various stages, introduce major or minor drawing changes, e.g. amending a size, adding missing information, change to suit manufacturing constraints. The is done quite easily due to the fact that you are the only one responsible for all project information. However, in an engineering company, formal changes must be requested and actioned through formal change procedures. Every time this is done it costs the company money. Therefore, if Design can get the information as accurate as possible before drawing release, major cost savings can be achieved and manufacturing efficiency greatly enhanced. This is why the concurrent engineering approach referred to in Chapter 1 is so important. Once a sufficient knowledge base is obtained, EA projects can act as an application of this concept. Then, when you have completed them and the rest of your course, you will be able to work in companies using either the traditional or modern approach to product development.

In order to manage information release, a company usually has an administration configuration and control system through which drawing releases or changes-to-drawing requests are issued and actioned (Figure 2.3).

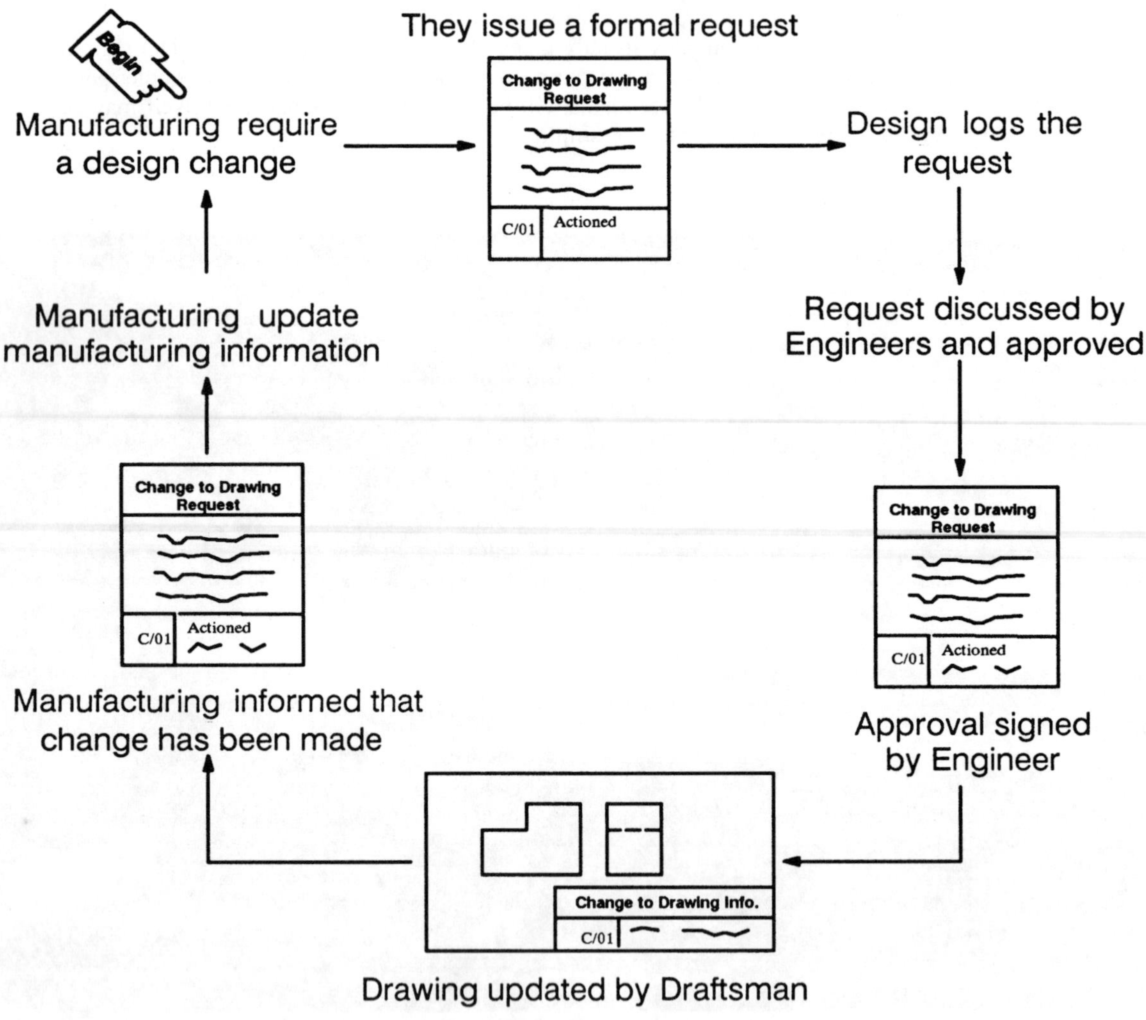


**Figure 2.3** *A simple engineering information configuration and control system*

As new drawings are released to Manufacturing, formal systems such as this inform them of the existence and the status of the product. Existing drawings may be changed as mentioned previously and such systems inform downstream departments that the new drawings are ready for use.

Other similar systems may exist for the communication of information between other departments, e.g. for the release of process planning information from Manufacturing to Production Planning.

When drawings are changed they will be updated with a revision number so that anyone in the company knows the latest design state of the product. The reason for each change may also be noted on the drawing or on a separate drawing change document, the number of which will be written on the drawing against the new revision number. However, this costs money in relation to time and effort. Concurrent Engineering tries to circumvent this.

## 2.1 The design function

The human race has always included those who are creative, intuitive, enquiring, methodical, analytical and rational. Many other characteristics can no doubt be applied to the species as well; however, those listed relate mainly to the requirements for the Design Engineer. In this list there are a number of uncomfortable contradictions caused mainly by the traditional engineering education approach which may confuse young people who decide to take up engineering as a career.

Most traditional engineering courses tend to concentrate on the analytical side of the subject; therefore most students entering such courses have, through schooling, deliberately taken subjects which have developed skills and thought processes based around the analytical approach to problem-solving.

It can therefore come as quite a shock to discover that a large proportion of product engineering, in particular design and process planning, is very subjective, with no clear problem definition and no one solution. This is certainly the case in industry.

Too many engineering subjects, and the way in which they are taught, give the impression that everything in engineering life is cut and dried. This most certainly not the case. Design, in particular, involves formulation of the problem, followed by synthesis and evaluation of possible solutions. Somewhere along the line design decisions and solutions involve compromise.

The analytical skills developed need to be used as an important subset of tools to determine whether certain aspects of a design are possible or functional. However, the creative part of design involves a great deal of intuitive thinking and supposition.

Although the EA projects in this book are well specified such that a general product solution is obvious there are many areas where intuition and analysis are required. Not all decisions have been taken on your behalf. On the contrary, many 'blanks' will have to be filled in. It is due to this that a wide variety of design detail can be produced by a class of students from the same project specification.

Do not be intimidated by the need to be creative or subjective during the design process. Indeed, the less inhibited you are early on in the process the better, because more genuine alternative solutions to the same problem then become apparent.

Once confidence in generating alternative solutions is gained, designing can be a very enjoyable and worthwhile experience; do not focus too quickly on a solution but view and evaluate alternatives.

Prior to the Industrial Revolution the vast majority of the population lived in rural societies which were mainly agriculturally and craft based. The design of products tended not to be separate from their manufacture. For example, the blacksmith in making a horseshoe or chisel would probably manufacture the item from memory, without any formal drawings. Thus each item would tend to be unique with assembled parts, such as those in a mechanism, not being interchangeable.

As more complex products were required and more complex manufacturing processes developed, more people were required to engineer products. The design function was separated from manufacturing. Also actual production was separated from production planning and process planning. This is now one of the main features of modern manufacturing practice. Indeed, modern civilization would not have developed as it has if these activities had not been separated and organized to involve larger numbers of people working in larger organizations with people who specialize in one skill or a small range of skills.

It is ironic that the modern concurrent engineering approach to the development of products tries to obtain the benefits of the original craftsman's combination of skills by introducing a team approach which brings together all the specialities required. Hopefully, your EA experience will help you appreciate this.

## 2.2 The communication of design information

Many departments and employees in a company use the information produced at the design stage of product development. It is the key information which drives product manufacturing and delivery.

As mentioned earlier, the main form in which finalized designs are communicated in an engineering environment is through the use of engineering drawings. However, other forms of drawing are usually created before the final formal engineering drawings are created and issued for manufacture.

Drawings, in many different forms, are used by all types of designers – not only engineers but also architects, product designers and graphic designers. They are used to develop ideas and concepts, to communicate these to other people and to aid the process of detailing manufacturing requirements. Drawings of all types are a communication language understandable by a wide range of people within and outside the company using them, e.g. customers, suppliers, subcontractors and workshop staff.

Designers use all forms of drawings to communicate information, the type and style of drawing used being dependent on the stage of development of the project and the requirements of other people involved in the product design and development process. Drawings do not necessarily imply paper documents; they could also consist of computer-aided design data and information. For the purpose of EA projects the following types of drawing are recommended for the development, presentation and communication of your ideas:

- sketches
- layout drawings
- assembly or general arrangement drawings
- detailed part drawings.

These are listed in the order in which they are usually created during the design process and are the most common types of drawings used during the design stage of engineering a product. In some cases you may decide to miss out the layout drawing stage and go on to the assembly and/or detail drawing stage depending on the complexity of the idea you are trying to develop. To help you understand the role of each type of drawing we will now describe examples of each.

### 2.2.1 Sketches

Sketching is the earliest phase of the design process, free-hand paper sketches usually dominating at this stage; the use of CAD systems at this phase is rare. This activity is usually carried out after a product specification has been produced which outlines the market/customer needs, physical limitations, operational environment, etc. of the product and records, in hard copy, the ideas taking shape in the designer's mind. Usually the designer will not go into any great detail but will allow his or her mind to wander free generating as many different ideas and concepts as possible; developing a number of potential solutions in the process. The initial sketches will tend to focus on the overall system and its operation, helping to develop different ideas.

There are no formal rules for sketching and usually the drawings are not drawn to scale. Their role is to develop concepts. This is illustrated in the rough sketch for the machine vice given in Figure 2.4. It is not to scale, it is not drawn with any particular flair or artistic merit and some of the detail is missing (this has been picked up in more formal drawings as will be shown later). However, it is adequate to demonstrate a solution, to promote a discussion between the designer and other people involved in the product development process and to focus the mind on potential problems and other considerations related to the product, e.g. manufacturing process requirements, raw materials. The important thing to remember is that as soon a sketch or sketches begins to take shape other alternatives, improvements and contributions can be forthcoming. The designers should not be totally inhibited at this stage by constraints on the product, e.g. by available manufacturing processes or raw materials; they should initially concentrate only on the design solution itself. *The most important thing at this stage is to get ideas down on paper*! Do not worry about the difficult parts of the design: these will be addressed later in the process. Indeed, sketching will also trigger ideas about more difficult problems and will also highlight parts and problems which may require some analysis.

Some rough calculations at the sketching stage will enable important decisions to be taken, e.g. which materials will be used, what the product's overall operational capability will be. As you become an experienced Design Engineer you will develop a 'feel' for materials and processes and, using a few calculations, will be able to estimate with reasonable accuracy sizes and cross-sections of key components, supports, etc.

'Gut feeling' also has a role to play in the design process. Many of the solutions developed will be subjective ones – do not worry about or be uncomfortable with this; as mentioned earlier it is very much part of the engineering process.

As your sketches develop, you may discover that the proposed solution will not work and you may have to go back and modify the solution in line with new data. This is a very natural part of the design activity and part of the learning process. You should not look on this as being failure. Also, be prepared to accept constructive criticism, do not take it personally. Compromise, discussion and flexibility are crucial at this stage.

### 2.2.2 Layout drawings

Eventually, freehand sketching will give insufficient detail to allow the project to progress as more accurate sizes are required and more detailed analysis needs to be carried out. The concept needs to be developed

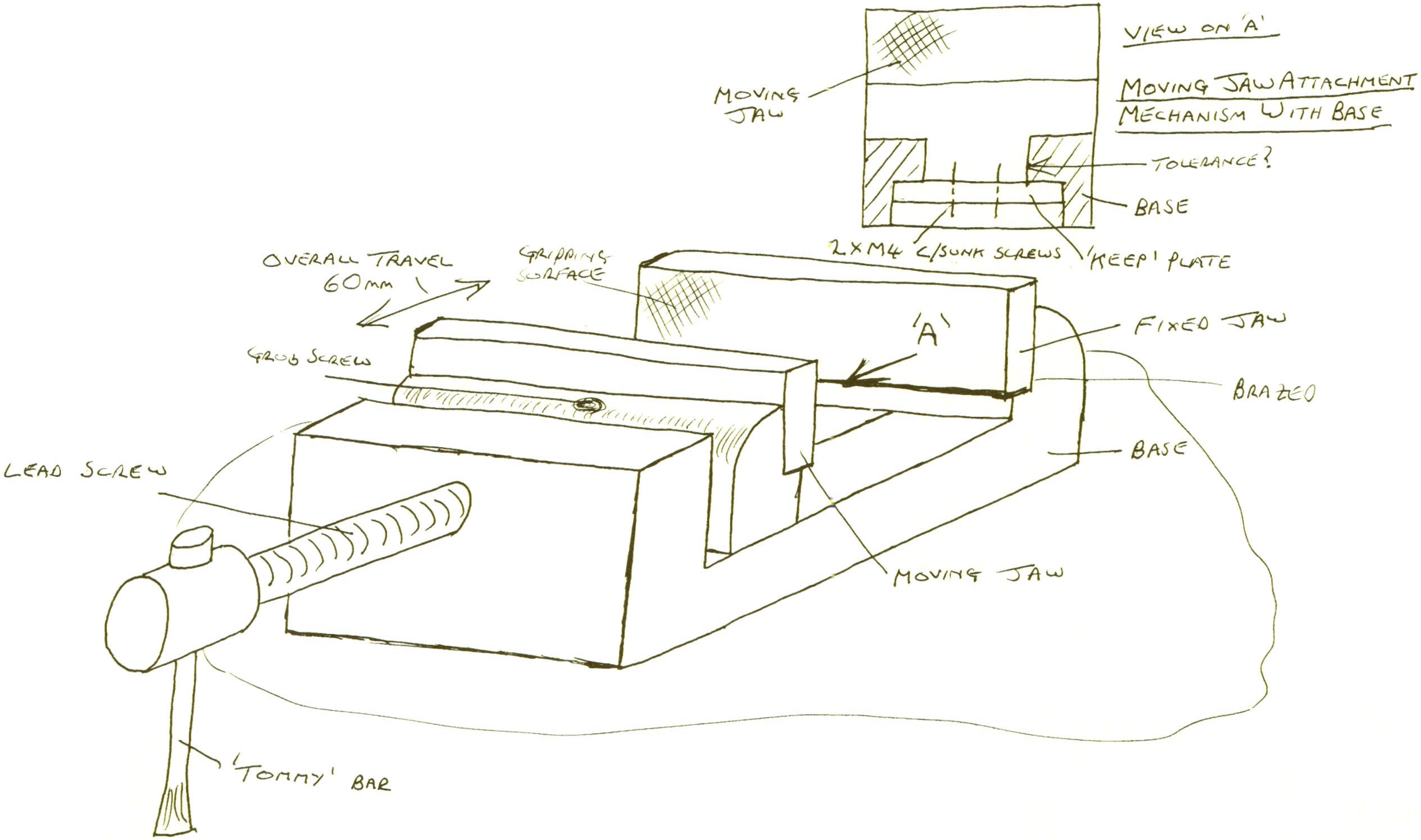


**Figure 2.4** *A typical sketch*

further and this phase is carried out using *layout drawings*. During the creation of these, components, subsystems and subassemblies can be analysed for cost, durability, performance, reliability and ease of manufacture. The designer, using information from a variety of sources, consultations, calculations, catalogues, codes of practice and standards will establish the sizes and types of key components, dimensions, tolerances, finishes and functional requirements.

The final layout drawings are drawn to scale and are used to check that parts will fit together. They also help confirm that the fundamental concept will work within the limits specified on the original specification.

Usually at this stage only the key features are drawn, as shown in Figure 2.5 – not all details are specified and only key dimensions are shown. Transparent overlays can be used to check the positions of key moving parts or assemblies and to determine movement envelopes. These drawings are the best method you can use to check for part interference when moving parts are involved. Other views may be used to clarify specific features of the design, e.g. orthographic projections, key sections. Usually all three views are developed simultaneously so that the effect of a design change in one view can be observed in another, e.g. for interference. Components are laid out to determine the relationships between parts and provide information on the structural elements that will support and/or house them. CAD systems may be used extensively at this point.

At this stage, detail manufacturing requirements can also be judged – the placing of components within the design on key axes, e.g. horizontal or vertical, may enhance manufacturability. However, restraints such as these should not be overriding and should be discussed with the Manufacturing Department.

Another key decision made at this stage will be whether the parts, subassemblies or final product will be purchased from a supplier or made in-house. Raw material and supplied or proprietary items will also be chosen depending on past company practice. Purchased items, such as bolts, nuts, flanges, electric motors and bearings, need to be looked at carefully, considering not only their performance and size but also cost, availability, etc.

Layout drawings do not contain all the information required to enable all the detail parts, subassemblies and the final assembly to be manufactured. This is where the final two types of drawing, namely assembly drawings and detail drawings, have parts to play.

### 2.2.3 Assembly drawings and detail part drawings

Formal engineering drawings, fully dimensioned and specified, are the main output from the engineering design process. Their main purposes are:

- to communicate the information required to enable the creation of manufacturing assembly and detail part process plans
- to communicate the design requirements to the shop floor and suppliers
- to convey to the customer the product form.

It is important that these drawings are produced in a format which is readily understood by all concerned. It is for this reason that a number of standards, both national and international, have been developed to aid engineers in the presentation of design information, e.g. BS308 [4]. The drawings arising from projects in this book should always be drawn to the appropriate national standard, in orthographic 3rd-angle projection.

A typical *assembly drawing* (also known as the general arrangement drawing) is shown in Figure 2.6. This shows the overall outline of the finished product assembly and the way in which the parts fit together. It gives the key overall dimensions, any limiting functional dimensions of moving parts and a parts list or bill-of-material. The latter, which may be drawn in a table on the original drawing or written on a separate sheet, lists all the subassemblies, manufactured detail parts and proprietary items which go into making the final assembly. The detail parts in the subassemblies are not mentioned in the parts list at this stage.

Many assemblies will require subassembly drawings which take the same form and specify the subassemblies and other parts associated with them, and it is here that their detail requirements are listed on another parts list. For example, a car assembly drawing will have subassembly drawings for the engine, suspension system, etc.; in turn the suspension system will have another subassembly drawing for the wheel assembly, and so on. Many hundreds of drawings and parts lists may have to exist to describe a product. This structure is illustrated in Figure 2.7.

These individual subassembly and detail part drawings show the detailed information relating to their design, performance and manufacture. These can show the design in more manageable sections and in more detail than can the assembly drawing.

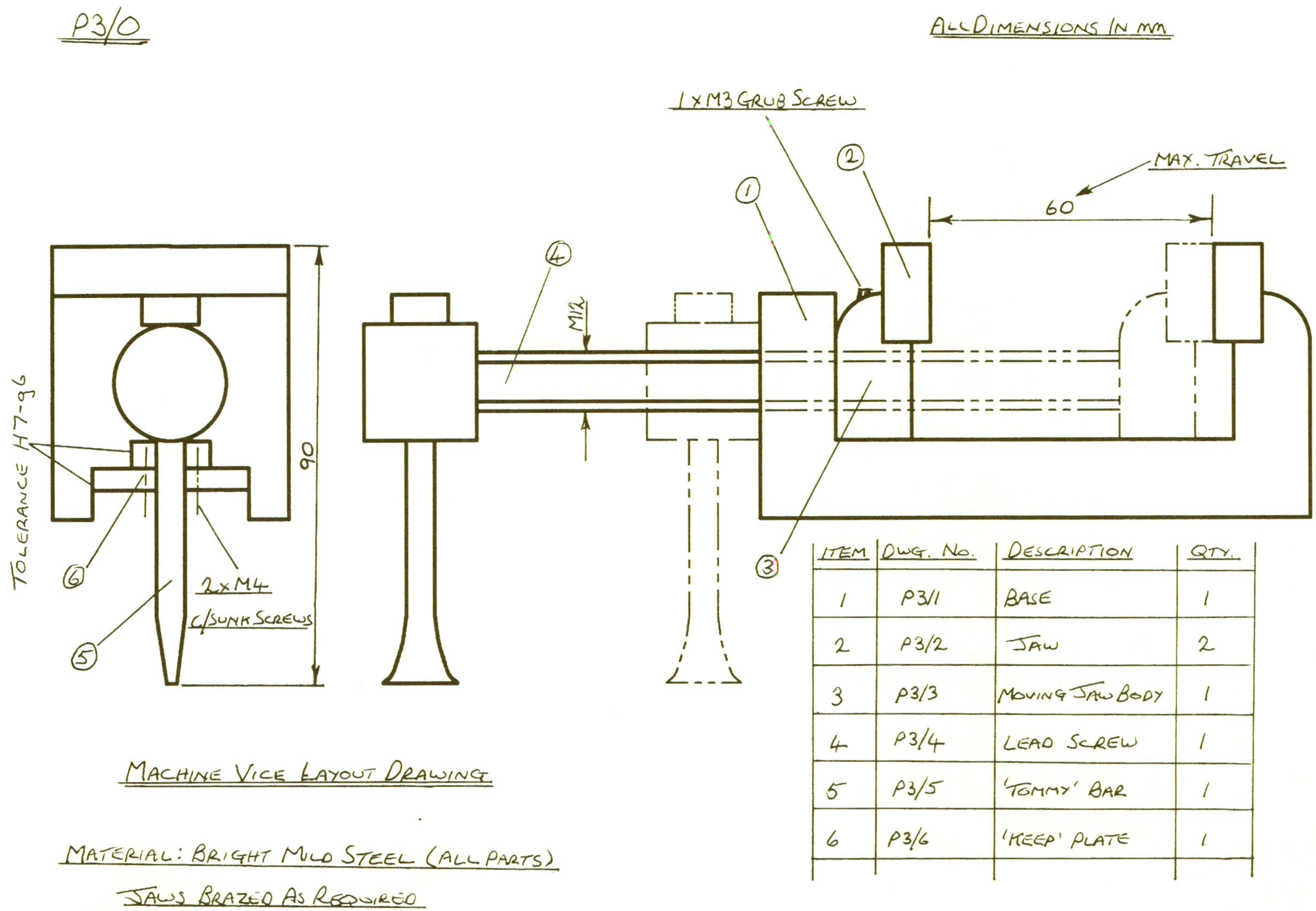


| ITEM | DWG. No. | DESCRIPTION | QTY. |
|---|---|---|---|
| 1 | P3/1 | BASE | 1 |
| 2 | P3/2 | JAW | 2 |
| 3 | P3/3 | MOVING JAW BODY | 1 |
| 4 | P3/4 | LEAD SCREW | 1 |
| 5 | P3/5 | 'TOMMY' BAR | 1 |
| 6 | P3/6 | 'KEEP' PLATE | 1 |

**Figure 2.5** *A typical layout drawing*

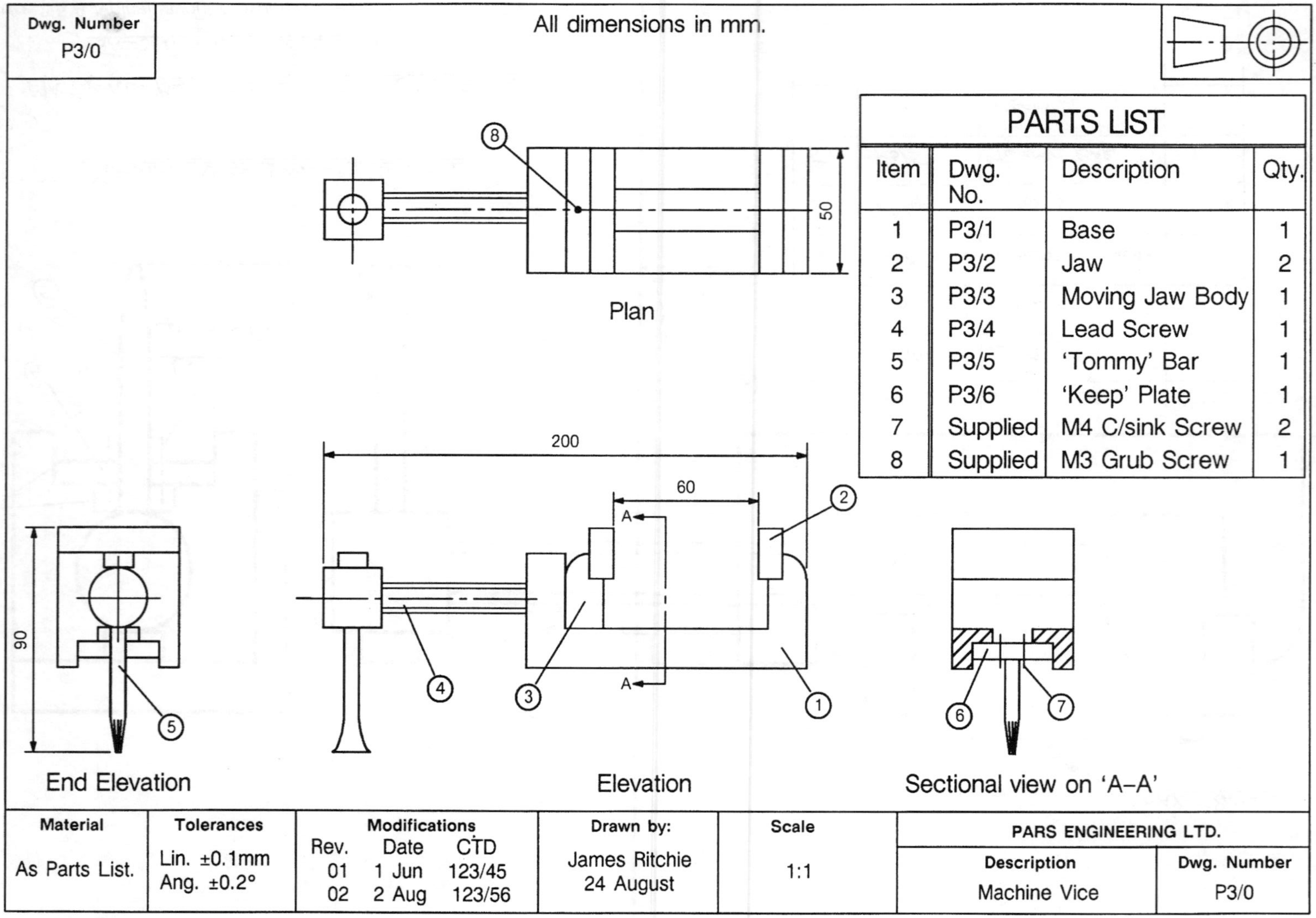


**Figure 2.6** *A typical assembly drawing*

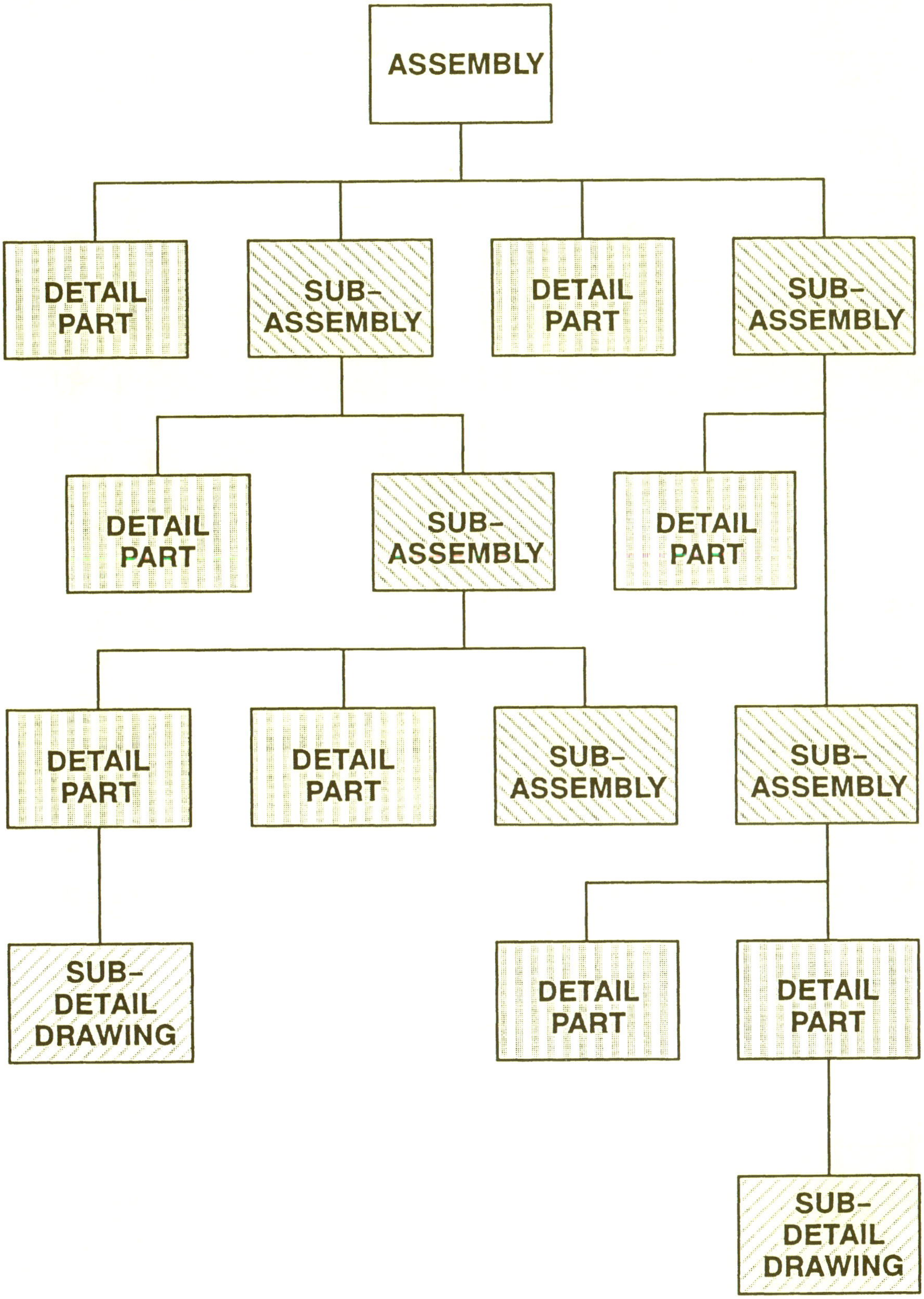


**Figure 2.7** *A typical product drawing structure*

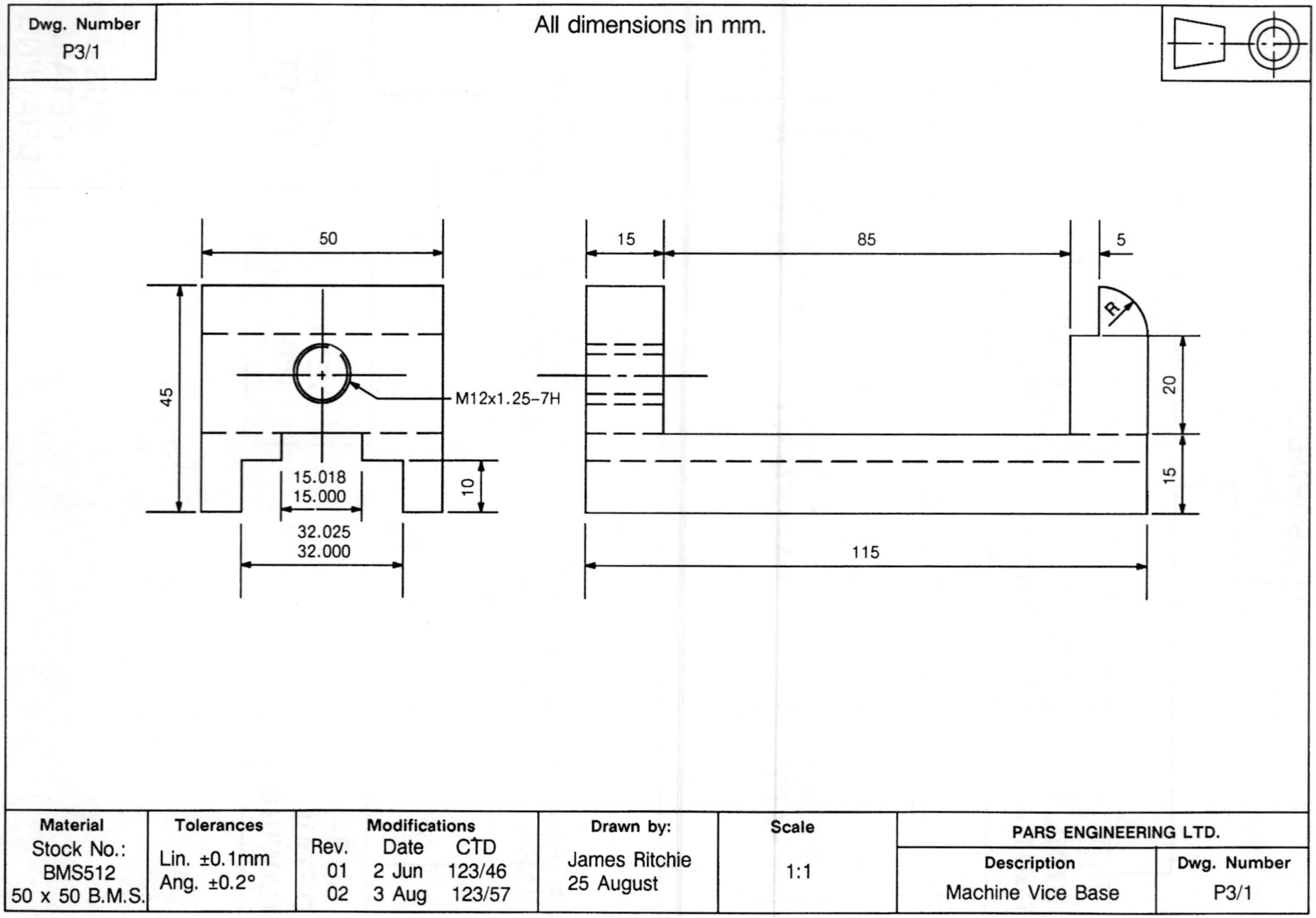


**Figure 2.8** *A typical detail drawing*

All individual components which need to be manufactured, either by a supplier or in-house, require a *detail drawing* (Figure 2.8). This gives dimensional and other geometrical information relating to the part. All dimensions must be specified, e.g. tolerances and finishes, as these dictate what manufacturing constraints exist and what manufacturing processes are required. They also state raw material requirements, heat treatments and other finishes, e.g. plating, painting. Tolerance and finishes have significant effects on manufacturing costs and must be chosen with great care. Note how much more detail is on Figures 2.6 and 2.8 than in the sketch shown in Figure 2.4.

Sometimes a detail drawing may require a *sub-detail drawing* which outlines a raw material form required for manufacture, e.g. a forging drawing so that the raw forging can be supplied in the form required for machining.

At the detailing stage many small but vitally important points about the design are decided and, ultimately, part fitness for purpose and performance are fixed by the quality of the information produced at this stage, e.g. dimensions shown, tolerances chosen. If the design is wrong, even in the most minute detail, this can have a catastrophic effect on performance and be very expensive, e.g. late design changes requiring major, expensive changes to manufacturing tooling or processes leading to poor customer perception of the company and a bad reputation. Also, these drawings are eventually used by process planners to produce manufacturing information as well as being used on the shop-floor for checking, etc. so good quality of information with no ambiguity is very important to the speeding-up of the production process.

## 2.3 Some notes for EA projects

Although the Design Department is usually responsible for developing product specifications, in the case of the EA projects in this book this has already been done. You will, no doubt, gain experience of generating customer design specifications in later design courses.

Once the specification for the design is read and understood the design process can begin. The recommended approach for developing solutions is given in Figure 2.9.

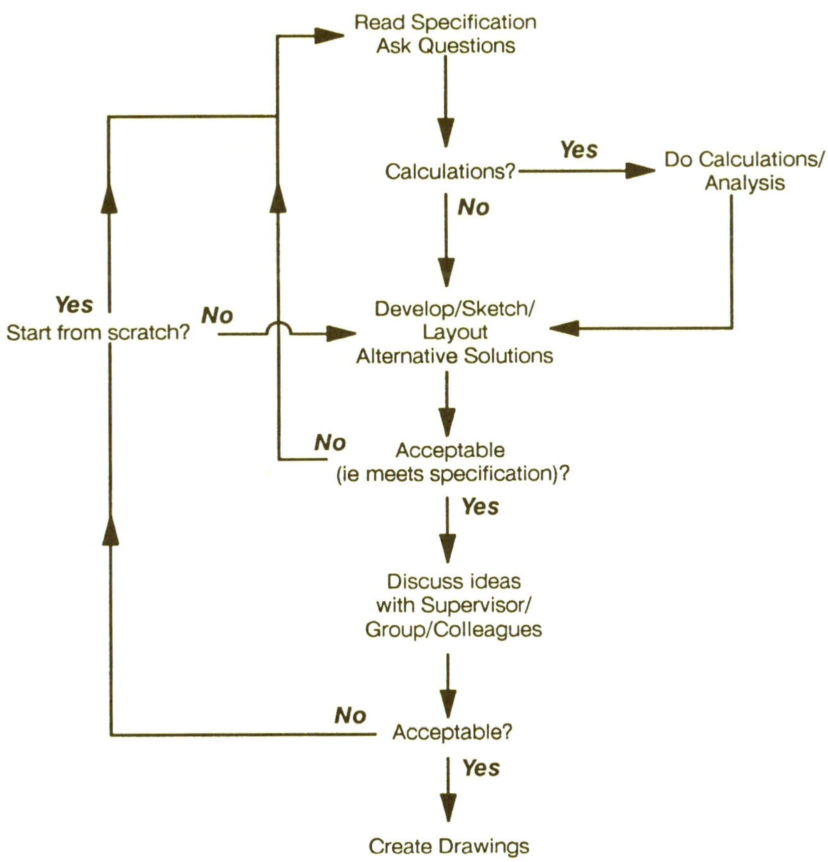


**Figure 2.9** *Recommended design strategy for EA projects*

It is important to remember that, from a design point of view, EA projects will give an appreciation of the design process, of where analysis fits in, of how manufacturing should be considered and of how to formally present finalised design information.

This grounding will lay the foundations for formal engineering design teaching including design methods, evaluation and strategies at a later stage of any engineering course. The EA projects deliberately do not allow design to overwhelm all the other aspects of engineering a product, but provide an awareness of the overall process of engineering a product and how each stage affects the others.

The simple design strategy given in this section is for EA projects of the type given in this book; it is not a formal design strategy for all engineering design.

All parts and assemblies that you design should be allocated a structured part number/drawing number, as is the case in industry. For example, 'P3–7' may be the part number for component 7 in project 3.

All assembly drawings should be balloon labelled and have a parts list which gives the drawing number and quantity of each part used in the assembly, including supplied items, e.g. nuts, bolts, screws. Only the overall assembly dimensions should be given; detail part sizes should only appear on detail drawings.

Each original detail part should be drawn on a separate piece of drawing paper or in a separate CAD file, labelled with its own unique part number and dimensioned according to the standard in use. This emphasizes that each individual part, suitably toleranced, can be made individually for a spare if required, as in industry. Raw material descriptions and cross-section or stock numbers (if any) should be specified as well.

On completion of each project you will have a well defined and structured drawing set; this, along with calculations and other project material, should be filed in a separate section of your EA logbook.

It is important to stress the need for good quality information. In industry, finished drawings may eventually be used to sell the product to a customer, therefore neatness and tidiness are of paramount importance. Regardless of how well engineered a product is, poor draftmanship can lose customers.

One excellent text which brings together much of the fundamental knowledge required, particularly for more complex engineering project design using fittings, belt drives, gears, etc. is reference [5].

Remember, ideas can be developed when they are put down on paper and concepts can be discussed. The production of good quality, unambiguous engineering drawings is something which should be second nature to any professional engineer.

## References

1 *Essentials of Engineering Design*, J. Walton, West Publishing Company (1991)
2 *The Mechanical Design Process*, David G. Ullman, McGraw-Hill (1997, 2nd edition)
3 *Mechanical Engineering Design*, J. E. Shigley, McGraw-Hill (1977)
4 *Engineering Drawing Practice for Schools and Colleges*, British Standards Institution (1986)
5 *The Student Engineer's Companion*, J. Carvill, Butterworth (1990)

# 3 *Preparing for manufacture*

Manufacturing processes can be classified into two distinct groups depending on their outputs, that is whether they produce discrete products or continuous products. Discrete products are products which are made up of individual items, such as nails, cutlery, bolts, or assemblies such as computers, televisions or engines. Continuous products are manufactured using continuous flow processes and may either be chemical products, such as petrols, oils, gases, etc., or raw materials produced for the eventual manufacture of discrete products, such as plastic tubing, sheet steel, wire, etc.

The design methods, manufacturing processes and projects referred to in this book relate to discrete products. Historically, when products to be manufactured became more complex, as did the manufacturing processes involved, it was obvious that no one person could carry out all the tasks required during the design and manufacturing stages. The skilled craftsman-based approach had to be replaced. Also, the volume of demand was so high that methods of manufacture had to change.

As factories evolved, so did the specialized skills of the people who worked in them. Each individual tended to carry out only one part of the manufacturing process, then pass the partially finished parts on to the next stage for the following manufacturing operation. Not only had the design function become separated from the manufacturing function but manufacture was broken up into the individual processes required to make or assemble each item.

One of the first examples of a modern factory was the one set up in the early 1800s by Eli Whitney, the inventor of the universal milling machine. Unfortunately, then as now, most of the developments of new technology and manufacturing methods took place through the requirements for more powerful and effective weapons. In Whitney's case he was asked by the United States Federal Arsenal to produce a large quantity of muskets. Prior to this each musket mechanism was manufactured and assembled by a skilled craftsman. This meant that each assembly was unique so parts from one assembly could not be used on another because they were individually crafted. Whitney developed a number of new machine tools to produce the detailed musket components. The actual assembly was broken down into simple discrete stages which were then carried out in a line flow environment by a team of people, i.e. divide the work and multiply the output. This meant that parts were designed and manufactured to specific tolerances and therefore they could be interchanged and spares made where required. This was one of the first examples of interchangeability and is fundamental to the development of modern-day products.

So why is good preparation for manufacture so important?

If raw materials, manufacturing resource requirements, manufacturing methods and shop floor personnel are well planned and organized prior to the production of a product in the factory then it should be feasible to make it as quickly as possible to meet market requirements.

Well structured stage-by-stage instructions created at the process planning stage for each part or assembly can reduce the amount of thinking time required for an operator to a minimum, thus allowing them to get on with the actual manufacturing operation. This good preparation also ensures that all the necessary tooling and other equipment is available to them when the part is to be made. Accurate and consistent operation times can also be prepared, enabling precise forecasting of resources and delivery schedules at the production planning stage.

Accurate production schedules, created at the production planning stage, properly detailing the raw materials required, manufacturing resources and purchased parts needed, as well as work priorities, can reduce delays and inventories such that significant cost reductions can be made. This, combined with good process planning and effective DFM, can give companies a crucial market advantage.

## 3.1 Types of production

The discrete product group of manufacturing processes can be further classified into mass and batch production.

Products are said to be mass produced if they are manufactured in very high volumes, either continuously or periodically over long periods of time. Usually the machines and processes used in this environment are organized in a flow line manner with each part or assembly passing through each one. Work loads are well balanced, with product demand stable and output independent of individual customer orders. Typical examples of products manufactured in this way are pens, nails, confectionery, washers, tacks, screws, chemical products, cars, lorries, springs. These are produced by specialized machines which tend not to be used for any other types of product manufacture.

Batch production quantities can vary substantially from 1 to 100,000 or more, depending on the manufacturing processes used and the product demand. Orders for these items are more likely to be based on actual customer demands, although many companies using this form of manufacture will produce for stock depending on market demand forecasts. The machines and processes used in this environment are usually more general in nature and can be used on a number of different product types. In this case the product may not be manufactured on all the resources available to the company, but only on selected work centres. The work load tends to be less well balanced than in the mass production environment and production management and control problems more complex. Typical products may be television sets, books, computers, flat-pack kitchen units, overhead projectors.

The people working in this type of production can range from being skilled to unskilled depending on the product's manufacturing requirements.

A specialized form of batch production is job lot production. Companies operating in this classification of manufacture usually produce very small batch quantities for specific customer orders which are not usually repeated. These usually vary from a batch or lot size of 1 to 600, although this range is not a limitation. Manufacturing processes applied are multipurpose and operated by skilled, well-trained personnel. Typical job lot industries include special purpose pumps, oil field Christmas tree assemblies, aeroplanes, ships, fitted kitchens, bank autotellers.

## 3.2 Manufacturing processes required for EA projects

The manufacturing processes which you should know about in order to carry out the projects in this book are listed in the logbook sheets supplied by your tutor. These logbook index sheets should be cut out and filled in as appropriate lectures, demonstrations and projects are completed. (Note that a blank grid is supplied for other topics, photocopy this sheet as required.) Unless predefined in the project specification, any of the processes listed should be available for the manufacture of project components. A detailed breakdown of each process and associated work holding devices and tooling should be given through specific lectures and demonstrations as part of your course. Many other supporting texts can also give this information [1,2,3,4].

Most of the projects involve the design and/or manufacture of either individual components or mechanical assemblies. Therefore all of the manufacturing processes involved are mechanical. It should be remembered, however, that the concepts experienced relate to all manufacturing sectors, e.g. electrical/electronic, chemical. The documentation produced will also exist in other forms in these industries.

## References.

1 *Manufacturing Technology – Volume 1*, R. L. Timings, Longman (1992, 2nd edition)
2 *Manufacturing Technology – Volume 2*, R. L. Timings, Longman (1993)
3 *The Student Engineer's Companion*, J. Carvill, Butterworth (1990)
4 *Manufacturing Engineering and Technology*, S. Kalpakjian, Addison-Wesley (1989)

# 4
# *Process planning*

Process planning has already been discussed in Chapter 1 when considering the function of the Manufacturing Department. All manufactured products and constituent parts must be process planned at some stage in their development in order to be manufactured [1,2]. This may also be known as process design.

The outputs produced during this process are described in more detail in this chapter.

The person who carries out this function within a manufacturing company is usually called a process planner.

This task is carried out with regard to the following:

1 how best to make the individual parts or assemblies with the resources available (e.g. machines, tooling, material, labour)
2 the machining, finishing and/or assembly requirements with particular reference to the drawing.

The information prepared at this stage may consist of some or all of the following:

1 raw material requirements
2 assembly lists (picking lists)
3 operations process planning (see later)
4 operation set-up sheets – these are graphical sheets which illustrate to the operator the details of the machine or process set-up and key manufacturing details
5 numerical control (NC) programs for controlling computer-controlled machine tools
6 tooling lists
7 tooling, fixturing or jig designs (see later)
8 manufacturing time estimates.

All details relating to part, product or assembly manufacturing requirements and the corresponding resource sequence must be outlined.

On receipt of a Works Order and its subsequent scheduling in a batch manufacturing environment, the following are passed onto the shop-floor as a package:

1 the drawing
2 process planning sheets
3 set-up sheets
4 job cards and/or travellers.

Figures 4.1 and 4.2 illustrate part of a professionally produced process plan and its corresponding set-up sheets; this was developed on a computer-aided process planning (CAPP) system. The other documents mentioned above have already been covered in full in Chapter 1 of this book.

Machining times for specific types of manufacturing operation will be calculated for some EA projects. This will give you experience of how these can be determined and used as a means of optimizing designs suited to manufacture and of calculating shop floor loading, bonuses, delivery times, etc. The times you calculate will also be compared with those achieved while carrying out the actual manufacturing process.

For machined mechanical piece parts, such as those made for EA projects, a process plan may be defined as: *a document which details the sequence of machining operations required to manufacture an item from raw material to its finished state as specified by the drawing.*

Note that plans can also be created for assembled or fabricated items. You may also be asked to prepare these by your tutor.

The method of manufacture chosen is dependent on a number of factors as outlined on page 7.

In modern industry it is normal for more than one individual and/or machine to be involved in the manufacture of any single part, therefore it is necessary to lay out detailed instructions for part manufacture at each stage. Other benefits will become obvious at a later date, e.g. production scheduling, ease of manufacture (operator not having to think about how job should be made, therefore manufacturing time not wasted), etc.

In the case of EA parts only one operator will be involved with a number of machine tools and processes – *you*. This is a unique situation and far from the norm. Any process plans created will assume that they are for manufacture in an industrial environment.

The operator is not simply handed a drawing and told 'make it'. Rather, the process plan is usually created some time before manufacturing begins. Indeed, it should not be possible to start manufacture without it.

Therefore a number of factors can and should be determined beforehand:

1 raw material dimensions can be fixed
2 machine tool requirements and capabilities can be assessed
3 an accurate estimate of manufacturing times can be made; only a few individual operation machining times will be determined for EA projects
4 jigs, fixtures, special tools and special gauges can be designed and manufactured.

*Jigs* and *fixtures* (Figures 4.3 and 4.4) are provided to convert standard machine tools into specialized machine tools[1]. They are usually associated with large-scale batch production and semiskilled operators. They are also used by skilled machinists when work is difficult to locate. *Jigs* may locate the workpiece and guide cutting tools onto it, e.g. drill jigs (Figure 4.3). *Fixtures* hold the workpiece and are fixed to the machine tool table but do not guide tools although they often incorporate a tool-setting device. Other types of jig can be designed for use in the production of mechanical assemblies or fabricated assemblies. These help speed up production considerably.

Special tools, jigs and fixtures are usually ordered, designed and made in a Tool Room, which is a form of jobbing shop.

The amount of detail on a plan depends on the type of manufacturing facility involved and the quality of operators; e.g. jobbing shop, small tool room, production shop; skilled, semiskilled, unskilled.

### Example

Assume the component shown in Figure 4.5 is to be made in:

(a) a jobbing shop with skilled operators
(b) a production shop with semiskilled labour.

Outline the first turning operation instructions on a centre lathe for each type of production facility. Assume the material is 50 mm diameter mild steel bar, 60 mm long.

| PLANNING | ITEM DESCRIPTION | DRAWING NUMBER | ISSUE |
|---|---|---|---|
| | Outer case (casting) | EX747493 | 01 |

| PRODUCT | PLANNING DEPT. | PLANNER | DATE |
|---|---|---|---|
| Fighter GD4 | Radar | A. Johnstone | 29 July |

| | | | |
|---|---|---|---|
| **MATL. CODE:** | 112112–12321 | **GT CELL:** | 'A' |
| **MATL. DESC.:** | Aluminium alloy casting | | |
| **MATL. QTY.:** | 1 per component | **METHOD.:** | P |

| PLANNING SHEET NO. | ISS. | PLANNING SHEET NO. | ISS. | SOURCE TAPES | NUMERICAL CONTROL TAPES | ISS. | SOURCE TAPES | NUMERICAL CONTROL TAPES | ISS. |
|---|---|---|---|---|---|---|---|---|---|
| 1 | 01 | | | S10 | EX747493T10 | 01 | | | |
| 2A | 01 | | | S20 | EX747493T20 | 01 | | | |
| 2B | 01 | | | S30 | EX747493T30 | 01 | | | |
| 2C | 01 | | | | | | | | |
| 2D | 02 | | | | | | | | |
| 2E | 03 | | | | | | | | |
| 3 | 01 | | | | | | | | |

| PLANNING SHEET NO. | ISS. | PLANNING SHEET NO. | ISS. | TOOL NUMBER | TOOL DESCRIPTION | OP. NO. |
|---|---|---|---|---|---|---|
| 4A | 02 | | | | | |
| 4B | 01 | | | LJ6824 | Milling fixture | 20 |
| 5 | 01 | | | SJ86795 | 12.5mm x 15 deg. angled slot drill | 20 |
| 6 | 01 | | | SJ88754 | 15mm special end mill | 20 |
| 7 | 01 | | | SJ88868 | 30mm dia. extended slot drill | 20 |
| 8 | 01 | | | MJ67784/1 | 30mm diameter Woodruff cutter | 20 |
| 9 | 01 | | | MJ67784/2 | 30mm diameter Woodruff cutter | 20 |
| | | | | MJ687784/3 | 30mm diameter Woodruff cutter | 20 |
| | | | | SJ100398 | Side and face cutter | 20 |
| | | | | SJ100546 | Special slot drill | 20 |
| | | | | MJ67225 | Toolholder for SJ100546 | 20 |
| | | | | LJ6929 | Fixture | 40 |

| TOTAL NUMBER OF SHEETS IN THIS PLANNING SET | 14 | PACKAGING INSTRUCTIONS | OP. NO. |
|---|---|---|---|
| | | | |
| | | | |
| | | | |
| | | | |

| | | | | | |
|---|---|---|---|---|---|
| **DRAWING NUMBER:** | EX747493 | **SHEET:** | 1 | **ISSUE:** | 01 |

**Figure 4.1** *CAPP system process plan*

| PLANNING OPERATIONS SHEET | | | | |
|---|---|---|---|---|
| **OPERATION NUMBER** | **CELL** | **MACH. CODE** | **OPERATION DETAILS** | **TAPE/TOOL DESCRIPTION** |
| 10 | X | 9824 | PROCURE MATERIAL FROM CUSTOMER'S PRIVATE | PRIVATE STORE |
| | | | STORE. CASTING NUMBER: 112112-12321. | |
| | | | | |
| | | | | |
| 20 | A | 6450 | LOAD 100MM SUB-TABLE | KTM 400 CNC |
| | | | SET UP AS PLANNING SHEET 3 | EX747493T10 ISS. 1 |
| | | | MACHINE AS PLANNING SHEET 4 | LJ6824 |
| | | | | MILLING FIXTURE |
| | | | ***** STRATEGIC CHECK ***** | SJ86795 |
| | | | DIMENSION DESCRIPTION DRW. GRID REF. | 12.5MM S.D. |
| | | | 201.66 OVERALL LENGTH L4 | SJ88754 |
| | | | 20.32 TOP FACE TO 1ST LEVEL K6/7 | 15MM SPEC. E.M. |
| | | | 96.20 RADIUS AT CENTRE J12 | SJ88868 |
| | | | | 30MM EXT. S.D. |
| | | | NOTE: | MJ67784/1 & 2 & 3 |
| | | | 'K' HOLE MACHINED TO 27.00 +0.01/-0.00 DIA. | SERIES WOODRUFF |
| | | | LARGE BORE MACHINED TO 7.080 +0.01/-0.00 DIA. | CUTTERS |
| | | | | SJ100398 |
| | | | | SIDE & FACE CUT. |
| | | | | SJ100546 |
| | | | | SPEC. S.D |
| | | | | MJ67225 |
| | | | | TOOLHOLDER |
| | | | | FOR SJ100546 |
| | | | | |
| | | | | |
| 30 | A | 9851 | DEBURR | DEBURRING BENCH |
| | | | | |
| | | | | |
| **DRAWING NUMBER:** EX747493 | | | **SHEET:** 2A | **ISSUE:** 01 |

**Figure 4.2** *Corresponding set-up sheets*

| PLANNING SET-UP SHEET | | | |
|---|---|---|---|
| DRWG. NO. EX747493 | | MACHINE 6450 – KTM 400 CNC | OP. NO. 20 |
| TAPE/TOOL DESCRIPTION | | OPERATION DETAILS | |
| LJ6824 | Milling fixture | Clock front edge of fixture square at front of machine table. | |
| Standard | 100mm Sub-table | Set up casting on fixture sitting on 3 tenons and locating with | |
| EX747493T10 | CNC Tape | datum lugs (2 positions) and at backstop. Apply pusher and | |
| | | supports. Clamp in 3 positions. | |

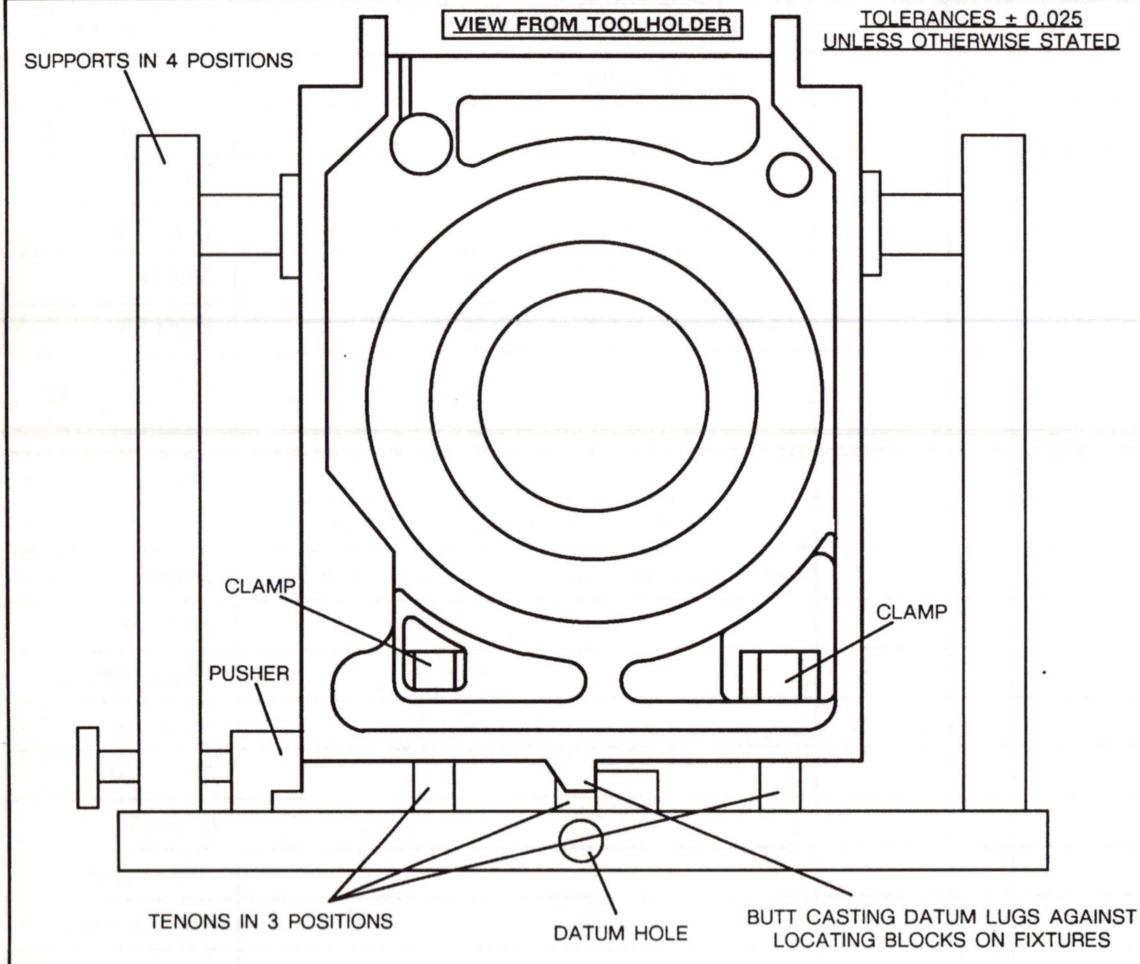


| | | | | | |
|---|---|---|---|---|---|
| | | | | | |
| | | | | | |
| | | | | | |
| **DESCRIPTION** | **GAUGE** | **LOW** | **HIGH** | **Drawn:** A. Johnstone | **Date:** 29 July |
| **DRAWING NUMBER:** EX747493 | | | | **SHEET:** 3 | **ISSUE:** 01 |

**Figure 4.2** *Continued*

## PLANNING TOOLING LIST

**JOB NUMBER:** EX747493 **ISS:**01 **CREATED:** 29 JULY
**OPERATION NUMBER:** 20 **MACHINE:** 6450 – KTM 400 CNC
**MATERIAL:** AL. AL. CAST **PLANNER:** A. JOHNSTONE

**XY DATUM:** CLOCK DATUM HOLE IN FIXTURE

**Z DATUM:** SET ALL TOOLS 75MM FROM THE FRONT FACE OF THE FIXTURE

**MINIMUM PACKING THICKNESS:** 1.5MM

**MAX. Z MOVEMENT:** 256MM

**X LIMITS:** –77.6MM TO 114MM Y LIMITS: 0.0MM TO 19.5MM

**ESTIMATED RUN TIME:** 183.9 MINUTES

| TLC. NO | TOOL NO. | BLOCK NO. | TOOL DESCRIPTION | TIME (MINS) | SPEED (RPM) |
|---|---|---|---|---|---|
| 1 | 1 | 30 | 20MM DIA. CARBIDE LONG SERIES SLOT DRILL | 5.57 | 2750 |
| 2 | 2 | 375 | 12MM DIA. LONG SERIES SLOT DRILL (LSSD) | 8.25 | 3438 |
| 3 | 3 | 1135 | 12MM DIA. X 15 DEG. LSSD SJ86795 | 4.82 | 1800 |
| 4 | 4 | 1205 | 16MM DIA. END MILL SJ88754 | 1.57 | 560 |
| 5 | 5 | 1280 | 6.5MM DIA. LONG SERIES SLOT DRILL (LSSD) | 4.43 | 3600 |
| 6 | 6 | 1855 | 5MM X 1.5MM RAD. LSSD | 4.49 | 3600 |
| 7 | 7 | 2340 | 3MM DIA. LONG SERIES SLOT DRILL (LSSD) | 2.21 | 3600 |
| 10 | 8 | 2590 | 28.5MM DIA. EXTENDED SLOT DRILL SJ88868 | 9.44 | 400 |
| 11 | 9 | 3010 | SPECIAL SLOT DRILL SJ100546 WITH TOOLHOLDER MJ67225 | 5.79 | 200 |
| 12 | 10 | 3070 | 30MM DIA. WOODRUFF CUTTER MJ67784/1 | 4.27 | 560 |
| 13 | 11 | 3295 | 30MM DIA. WOODRUFF CUTTER MJ67784/2 | 8.19 | 560 |
| 14 | 12 | 3725 | 30MM DIA. WOODRUFF CUTTER MJ67784/3 | 15.11 | 560 |
| 15 | 13 | 4540 | 100MM SIDE AND FACE CUTTER SJ100398 | 15.49 | 135 |
| OPTIONAL PLANNED STOP | | | | | |
| 15 | 13 | 4615 | 100MM SIDE AND FACE CUTTER SJ100398 | 19.97 | 135 |
| OPTIONAL PLANNED STOP | | | | | |
| 15 | 13 | 4700 | 100MM SIDE AND FACE CUTTER SJ100398 | 19.97 | 135 |
| OPTIONAL PLANNED STOP | | | | | |
| 15 | 13 | 4785 | 100MM SIDE AND FACE CUTTER SJ100398 | 19.97 | 135 |
| OPTIONAL PLANNED STOP | | | | | |
| 15 | 13 | 4870 | 100MM SIDE AND FACE CUTTER SJ100398 | 15.26 | 135 |
| 16 | 14 | 4970 | NO. 1 CENTRE DRILL | 0.33 | 3600 |

**Figure 4.2** *Continued*

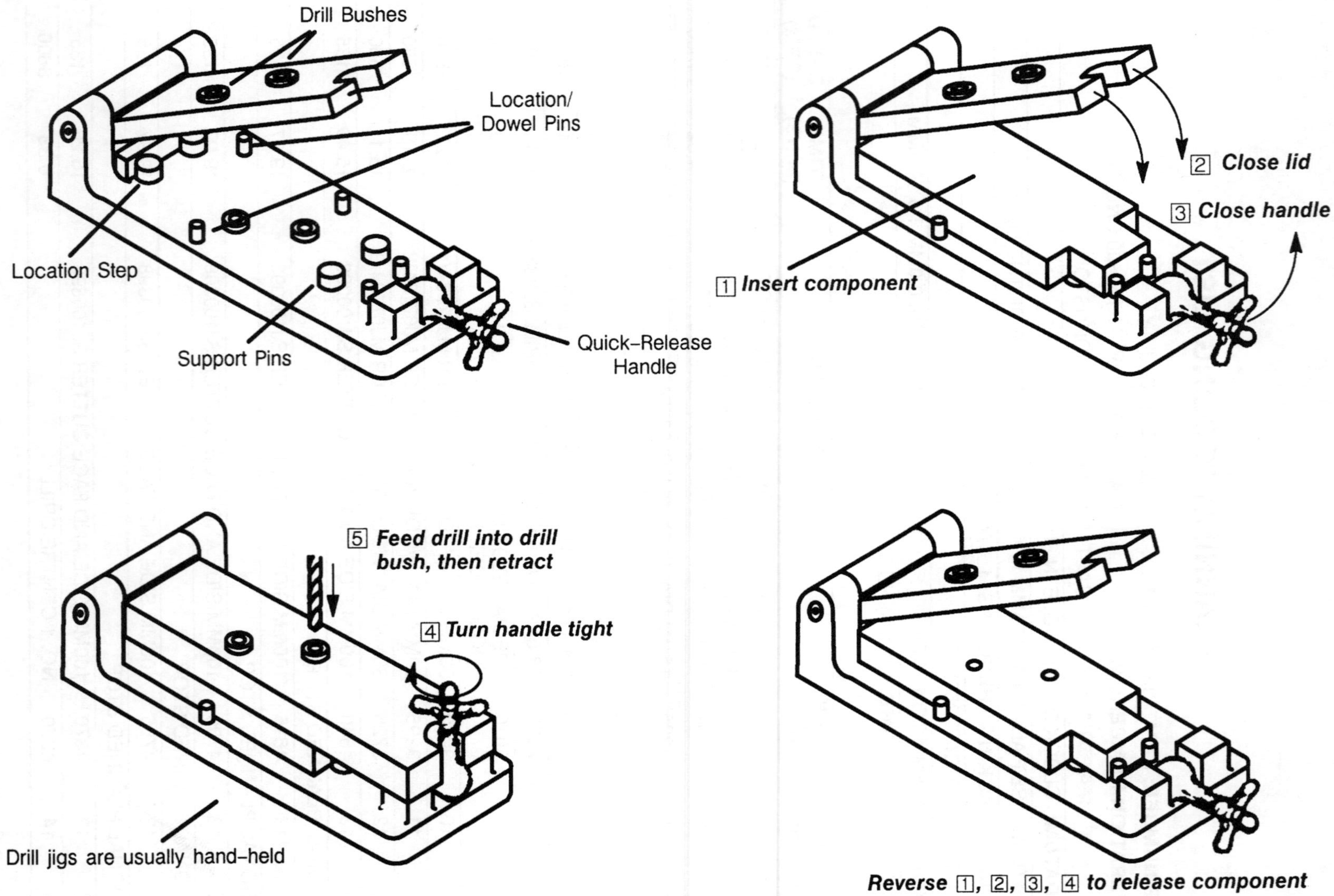


**Figure 4.3** *A typical drill jig*

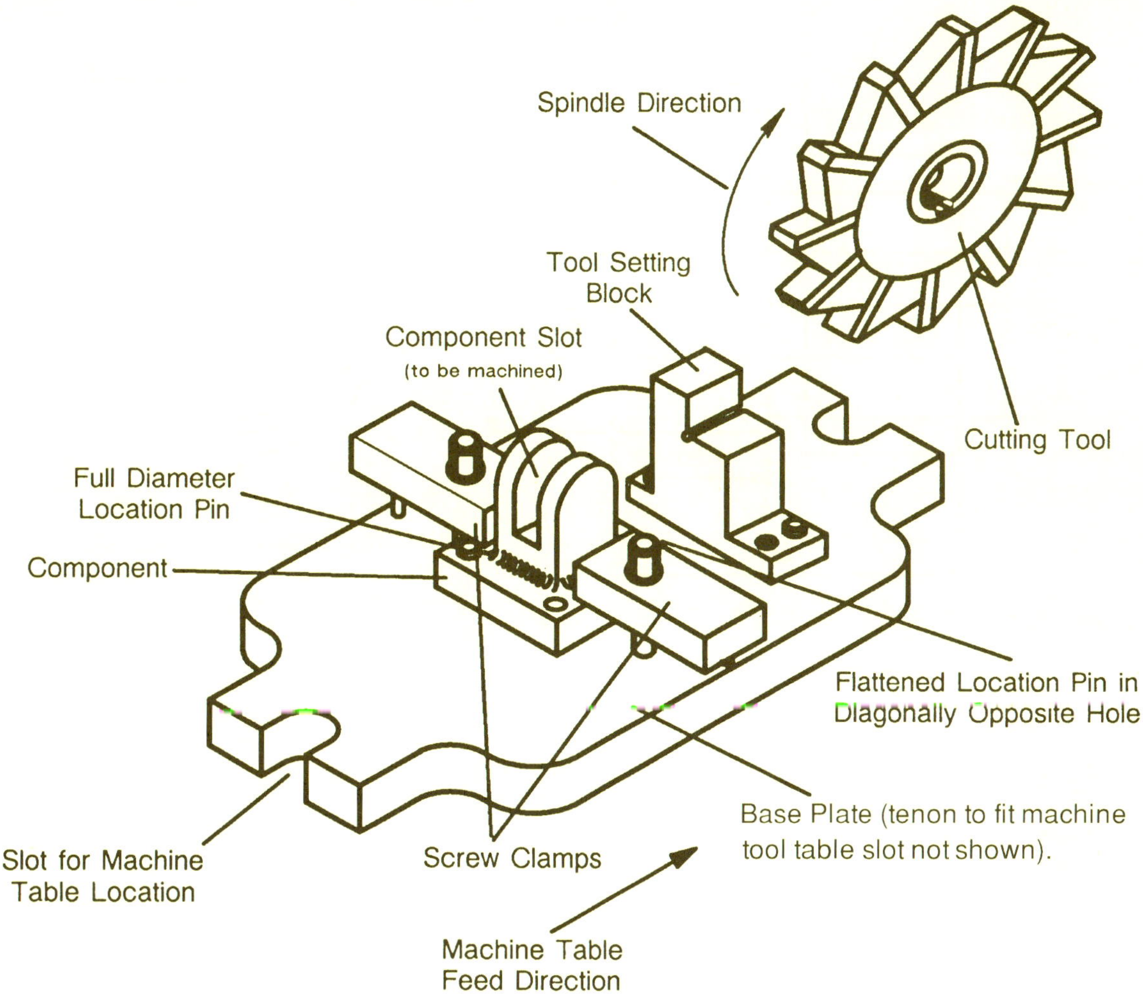


**Figure 4.4** *A typical fixture*

*(a) Jobbing shop*

Operation 10

1 Face bar
2 Turn external diameters as shown
3 Form undercut
4 Drill through hole
5 Part-off to length
6 Deburr and check.

*(b) Production shop*

Operation 10

1 Grip material in chuck
2 Face bar
3 Turn outside diameter
   Outside diameter 40 mm, diameter length 45 mm

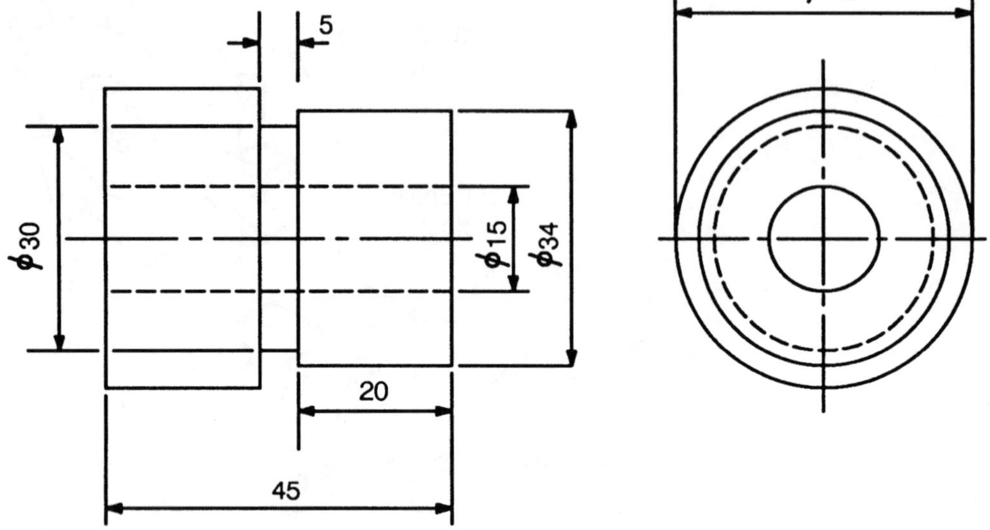


**Figure 4.5** *Planned part*

4 Turn 34 mm spigot diameter
   Diameter length 25 mm
5 Form undercut
   Undercut diameter 30 mm, width 5 mm
6 Centre drill end face
7 Drill 15 mm through hole
   Hole length 50 mm
8 Part-off to length
   Overall length 45 mm
9 Deburr and examine, inspect.

As shown, very basic instructions are given to the jobbing shop skilled operator, whereas the production shop operator must have detailed instructions. *These will include surface finishes and tolerances if the drawing quotes them.*

*Deburring*, which is usually carried out with a file or emery paper, removes sharp edges left by the machining process. This makes it safer for the next operator and other personnel to handle the components. It also provides clear edges if the surfaces concerned are to be used for location in a jig, fixture, etc. or as datums.

A simple planning operation such as this can be checked by making a rough sketch of the component and highlighting the machined surfaces in a different colour or thickness of pencil. This ensures that all the necessary machining has been carried out.

This technique may be useful for some of the components which you manufacture, especially those which are machined or fabricated.

## 4.1 Planning method

As a rough guide to the preparation of a process plan the following guidelines may be followed, as illustrated in Figure 4.6. This is an extension of the method suggested in (1).

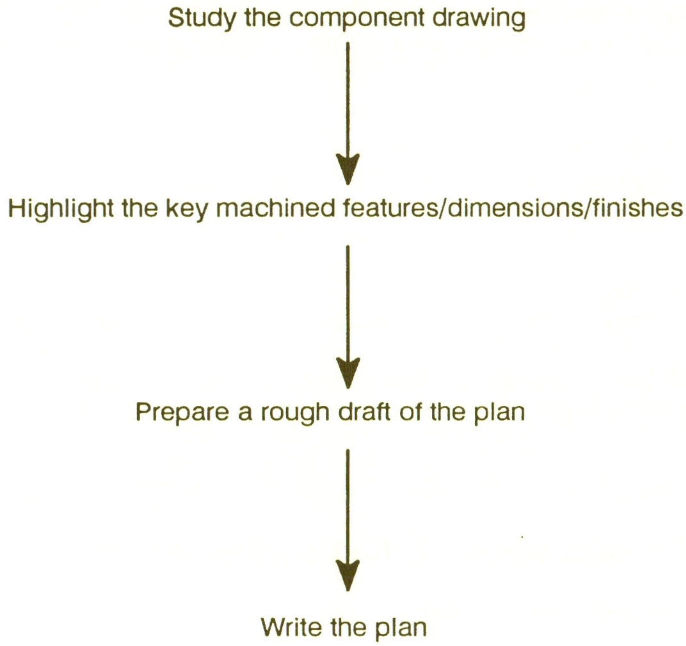


**Figure 4.6** *A process planning methodology*

The component drawing should be carefully studied:

1 so that important operational features can be determined
2 to check if the tolerances are for fits or clearances
3 to familiarize with size and weight of part. For example, is special lifting required?, which machines will it fit?, is a suitable machine available?

The main machined features should be highlighted in some way:

1 to indicate the areas to be machined
2 to help indicate if roughing or finishing will be required
3 to show important location and datum dimensions/features.

A rough draft or summary of the process plan should then be prepared, bearing in mind the following guidelines:

(a) **Establish at least one datum for machining**, e.g. a bar end face (see previous example), large flat surface for milling. A secondary datum may be a face at right angles to the main datum, or a bore.
(b) **Machine as much as possible from one setting**. Every new setting may reduce accuracy as datums have to be set again. In EA machining process plans a new set-up means a new operation.
(c) **Use the original datum as far as possible**. If it cannot be used any longer then new datums must have been machined accurately from the original ones.
(d) **Group similar operations together**, e.g. make all drilling operations follow each other. In a production shop this reduces machine handling times and assists the progress of work. The operator can also do all the operations on similar machines sequentially.
(e) **Perform accurate operations at the end of the manufacturing sequence**, e.g. grinding after turning or milling. Any potential damage to accurately machined or finished surfaces is minimized due to reduced handling. EA components will probably not require this.

(g) **Incorporate regular deburring operations** for reasons stated before.
(f) **Introduce key stage inspection operations**. This avoids unnecessary machining of already scrap components.

On EA project work, dimensions will be checked as parts are made. However, it may be useful to incorporate inspection operations or instructions in EA plans to highlight their importance. There may be a variety of different ways to machine a part and sequence the corresponding operations. The only way to determine the most economical method is to time and cost each one, unless an obviously ridiculous method has been employed.

The final process planning documentation can then be written. Blank process planning sheets will be supplied by your tutor; these should be photocopied for all EA process planning packages.

As previously suggested, all machining requirements can be checked using stage-by-stage sketches. Drawings may also call up special finishes, such as painting, plating, etc. – remember to include these details at the appropriate operations in the process plans.

## 4.2 Factors to consider when creating a process plan

When creating a process plan, a number of factors must be considered before producing the final document. The parameters covered are usually determined while 'highlighting machined dimensions' and 'making a rough draft'.

Although these factors have been listed separately, a number of them will be carried out in parallel:

(a) *Select material:* this will depend on the stock available and what is specified on the drawing as well as on what type of machining processes have been chosen, e.g. turned part or milled part. A stock list will be supplied by your tutor from which you will choose standard sizes of material. This list will also include standard store items such as nuts, bolts, screws, circlips, etc. Use the standard store item part numbers where required.

The raw material chosen for machined parts depends on their overall geometry and main external features, especially those to be produced during the first few manufacturing operations, for example:

- *external cylinders, tapers, etc.* will require round bar or tube
- *hexagonal, square or octagonal cross-section* but with external tapers, internal or external cylinders, etc. symmetrical about axis will need standard cross-section bar if available
- *box-shaped components*, comprising horizontal and perpendicular planes, slots, etc., require rectangular billets of varying dimensions; sometimes channel sections may be employed, depending on overall sizes
- *thin sheet components*, possibly comprising bends, where metal cutting is only necessary in two dimensions, usually require sheet metal
- *complex shapes* or *large components* may justify castings or forgings if reasonable production quantities are to be made.

Some items may be *assemblies*, therefore machined or bought parts have to be procured.

The above general rules do not always apply rigidly, especially if the grade of material is available but not in the desired form, e.g. only round bar when rectangular is required, no casting available but solid material is, etc.

(b) *Determine manufacturing or production sequence*: this is similar to what is carried out during material selection.

The raw material and the external geometry of the part usually determine the initial machining process which will carry out the main machining operations.

Further geometrical features such as holes, threads, slots, etc. along with tolerances and finishes also dictate what further machining may be required.

Some of this machining may be carried out on the same machine tool as the major material removal operations, e.g. external and internal surfaces machined on a lathe or milling of cut-outs and subsequent drilling of holes on the same face. Other features will have to be generated on other machine tools, e.g. a turned part requiring a slot or a hole perpendicular to the axis of rotation or a milled part requiring a hole to be jig-bored to a tight tolerance.

When the features to be machined are known and their subsequent machining operations determined then the *basic production sequence* is finalized. This is the order in which the manufacturing operations will be carried out and is called a plan summary. This is very similar to a job shop plan. For example:

| *Operation number* | *Machining process* |
|---|---|
| 10 | Cut material |
| 20 | Turn external and internal features |
| 30 | Reverse component and finish turn |
| 40 | Mill slot |
| 50 | Inspect |

Assume operations numbers are in steps of 10, since this makes it easier to insert later operations into a plan without renumbering.

As mentioned before, operations sequencing is carried out such that those on the same type or similar machine tools are grouped together. Sometimes tooling for operations will be similar, saving unnecessary storage of tools and fixtures.

(c) *Carry out individual operations analysis*: it is easier to plan the various operations in the order in which they have to be performed, treating each one separately but remembering their effect on following operations. For example, it may be that you are machining faces during one operation which will be used as datums in the next.

During phase (c), five main functions are carried out. Note that these relate mainly to machining but should also be considered for other processes.

### *1 Select the machine tool*

During the planning process and when considering the raw material or component state for the operation being planned, it is necessary to determine which machine tool is most suitable for the machining process chosen. It is best to choose the machine that is the best 'fit' and most suited to the batch size. *A list of the machines, processes and their associated capabilities should be supplied by your tutor.*

For example, a standard centre lathe may be chosen for a batch quantity of say 1–10 but a capstan lathe may be chosen for 10–200. Above this a more 'automatic' type lathe may be chosen.

In another example a jig boring machine capable of machining work of dimensions 2 m long by 2 m wide by 3 m high would not be chosen if the work external dimensions are only 100 mm × 50 mm × 80 mm and a smaller machine is available.

Machine tool accuracy is also a factor. The general process accuracy may dictate a boring machine to finish a tight tolerance hole rather than a drilling or milling machine. Also the machine tools available may be of the same type but different ages, therefore older machines may be less accurate.

Available machine feeds or, in the case of a lathe, gearing arrangement, leadscrew etc. may be other determining factors.

### *2 Select the cutting tool*

Cutting tools are selected and set up depending on the type of machining operations being carried out, the material being machined and the accuracy required.

- For *turning*, determine whether roughing and finishing is being carried out as well as other types of machining, e.g. facing, parallel turning, undercutting, forming.
- For *milling*, determine whether vertical or horizontal milling will be carried out. Tool diameter must also be a consideration as should depth of cut. If vertical milling is required, is it face milling, slotting, end milling, profile milling? If horizontal milling is required, is it face milling, slotting, gang milling? Tool types are chosen on this basis. All available tooling sizes should also be obtained from the stock list.
- For *shaping*, roughing or finishing or special form tools may be required.

In all cases, the possible *tool life* must be considered, particularly in a production environment and especially for different batch sizes. This in turn will dictate the tool material, e.g. tool steel, carbide tipped, high speed steel, although tools may be limited to those in stock.

The geometry of the tools concerned for different materials is dictated by standard tables for roughing and finishing tools in the manner illustrated in the appropriate section in Chapter 8.

The process plan only needs to specify special tooling, such as special cutters, jigs, fixtures, etc. It does not need to specify standard tooling, although references can be made to vice, chuck type and so on. However, in the case of EA project parts you should specify all standard tooling on the planning sheets.

### *3 Select the speeds and feeds*

Selecting the appropriate speeds and feeds is very important. These are determined from the tables in Chapter 8 and depend on the cutting tool and material to be cut. Examples of machining time calculations for different types of manufacturing process are also given in Chapter 8.

Again the final times are not usually quoted on the process plan but may be used at the estimating stage for determining the machining time of the part and overall lead time. This would be important in a production environment. You should compare these times with the actual times you achieve while carrying out an operation in your logbook. Explain the differences.

Detailed overall operations analysis would include the process of estimating the actual manufacturing time for the operation, including set-up time as well as machining time. Examples of typical calculations and standard set-up values for operation timing and costing are given in reference [3].

### *4 Select workpiece holding*

At every operation the process plan must mention the means of securely holding the workpiece. The work-holding device chosen depends on a number of factors:

1 the raw material or the partially machined component shape, e.g. thin walls, awkward profiles
2 the manufacturing process being used
3 the batch quantity
4 the accuracy required.

Usually for EA project turned parts a self-centring *3-jaw chuck* will be adequate. These are chosen if the workpiece is cylindrical or hexagonal and is to be machined about its central axis. Longer components will require a *centre* for tailstock support.

For smaller diameter workpieces, work-holding may be carried out using a *collet*.

Normal work-holding for conventionally milled or drilled components, due to their regular shape, will be on a *vice*. There will be no need for any special jigs or fixturing. If the component to be made is too large for a vice, but will still fit on the milling machine table, clamping directly to the table may be employed. Standard raw material sections will be used, no castings or forgings.

### *5 Specify inspection procedures*

*Workpiece measuring* is carried out using rules, calipers, combination sets and other metrology instruments. These methods of measuring and inspecting workpieces should be covered in metrology lectures and laboratories. On a process plan it may be necessary to specify specially made gauges (e.g. 'go' and 'no-go') for the operator's use. Standard measuring implements are not usually quoted. Special gauges will not be required in the case of EA projects.

Most of these instructions relate to machined parts. However, some products will be welded or assembled. Variations on these instructions should be applied when planning for these.

## 4.3 Process planning examples

Process plans will be written on formal process planning sheets.

A manufacturing planning package for each component consists of:

1 a planning front sheet
2 operation planning sheets – as many as are required
3 machining time calculation sheets (if required).

All front sheet information should be filled in, as should the individual operation details. The latter should contain operation number, stage-by-stage machining instructions, machine tools and tooling required.

Note:

1 pages should be numbered as shown in Example A below
2 1–3 lines of spacing should be left between operations so that additional operations can be inserted if required at a later date
3 operation numbers should be in steps of ten
4 always use block capitals.

Feed and speed ranges for the machine tools available to you should be supplied by your supervisor.

*Example A*
For the plug illustrated in Figure 4.7:

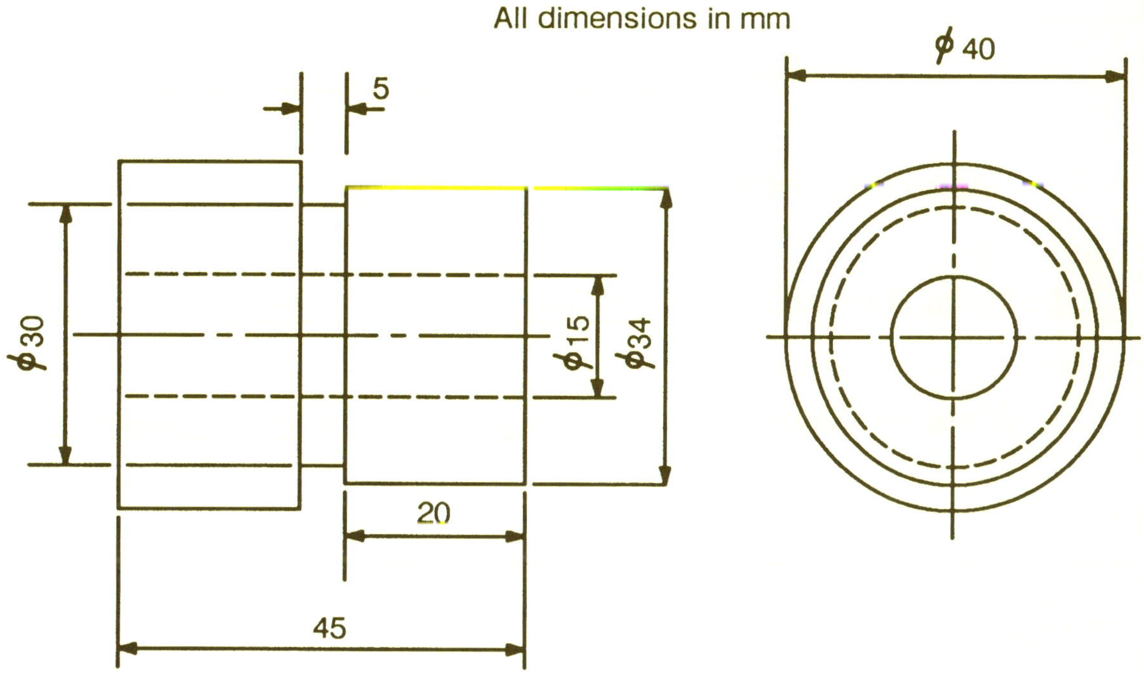


**Figure 4.7** *Plug*

1 prepare a manufacturing package
2 calculate the machining times required for the initial facing operation and turning of the external diameter.

Raw material 50 mm diameter mild steel to be cut in billets 80 mm long.
Only facing and plain turning machining times should be calculated.

| | |
|---|---|
| Feed = 0.25 mm/rev | Cutting speed = 25 m/min |
| Approach distance = 2 mm | Depth of roughing cut 2 mm |
| | Depth of finishing cut 1 mm |

*Assume for all machining calculations that the lathe speeds available are 46, 64, 100, 156, 289, 409, 639, 1000 revs/min.*

# ENGINEERING APPLICATIONS

# MANUFACTURING PLANNING PACKAGE

**Part Number–** EXAMPLE A

**Revision–** 1

**Part Description–** PLUG

**Material Description–** 50MM DIA. MILD STEEL BAR

**Material Quantity–** 80MM LONG = 1 COMPONENT

## Planner Details

**Name –** J. RITCHIE **Date –** 15 JUNE

**Class –** MECH. ENG. 1 **Project –** 4

**Page** 1 **of** 3

**ENGINEERING APPLICATIONS – Planning Sheet**

| Operation Number | Machine Description | Operation Description | Tooling Details/ CNC Reference |
|---|---|---|---|
| 10 | POWER | SAW MATERIAL TO LENGTH | MACHINE VICE |
| | HACKSAW | BILLET LENGTH 80MM | |
| | | | |
| | | | |
| 20 | CENTRE | GRIP MATERIAL IN CHUCK | 3-JAW CHUCK |
| | LATHE | 60MM PROJECTION | |
| | | | |
| | (a) | FACE BAR - REMOVE 1MM | FINISH FACING |
| | | | |
| | (b) | TURN OUTSIDE DIAMETER | ROUGHING PLAIN |
| | | DIAMETER 40MM | FINISH PLAIN |
| | | LENGTH 50MM | |
| | | | |
| | | TURN SPIGOT DIAMETER | ROUGHING PLAIN |
| | | DIAMETER 34MM | FINISH PLAIN |
| | | LENGTH 25MM | |
| | | | |
| | | FORM UNDERCUT | 5MM UNDERCUTTING |
| | | DIAMETER 30MM | TOOL |
| | | WIDTH 5MM | |
| | | | |
| | | CENTRE DRILL END | CENTRE DRILL |
| | | DEPTH 5MM | DRILL CHUCK |
| | | | |
| | | DRILL HOLE | 15MM DIA. TWIST |
| | | HOLE DIAMETER 15MM | DRILL |
| | | HOLE LENGTH 50MM | |
| | | | |
| | | PART-OFF TO LENGTH 45MM | PART-OFF TOOL |
| | | | |
| 30 | FITTING BENCH | DEBURR & EXAM | |

**Part Number –** EXAMPLE A  **Page** 2 **of** 3

## ENGINEERING APPLICATIONS – Planning Sheet

| Operation Number | Cutting Speed | | Cutting Feed | | Machining Time | |
|---|---|---|---|---|---|---|
| | Calculated | Actual | Calculated | Actual | Calculated | Actual |
| 20 (a) | 156rpm | | 0.25mm/rev | | 0.69min | |
| | | | | | | |
| (b) | 156 rpm | | 0.25mm/rev | | 4.00mins | |
| | | | | | | |
| | | | | | | |
| | | | | | | |
| | | | | | | |
| | | | | | | |
| | | | | | | |
| | | | | | | |
| | | | | | | |
| | | | | | | |
| | | | | | | |
| | | | | | | |
| | | | | | | |
| | | | | | | |
| | | | | | | |
| | | | | | | |
| | | | | | | |
| | | | | | | |
| | | | | | | |
| | | | | | | |
| | | | | | | |
| | | | | | | |
| | | | | | | |
| | | | | | | |
| | | | | | | |
| | | | | | | |
| | | | | | | |

**N.B. Remember to quote UNITS.**

**Part Number –** EXAMPLE A | **Page** 3 **of** 3

*Ex. A TYPICAL CALCULATION (see page 111 for schematic diagrams)*

*FROM TABLES*
- *Cutting feed* $f = 0.25$ mm/rev
- Cutting speed $S = 25$ *m/min*
- *Approach distance* $a = 2$ mm

$N$ = r.p.m.
$t_r$ = roughing machining time
$t_f$ = finishing machining time
$T$ = total travel (mm)
$L$ = length of cut (mm)
$t$ = total machining time (mins)
$D$ = diameter of work piece
$c$ = number of cuts

(a) *Facing* $D = 50\text{ mm},\ L = \frac{D}{2} = 25\text{ mm},\ c = 1$

$$T = L + a = 25 + 2 = 27\text{ mm}$$

$$\text{N} = \frac{1000S}{\pi D} = \frac{1000 \times 25}{\pi \times 50} = 159.15 \text{ r.p.m., say 156 r.p.m. from list (round down)}$$

$$t = \frac{cT}{fN} = \frac{1 \times 27}{0.25 \times 156} = \mathbf{0.69\ mins}$$

(b) *Plain turning* N.B. Diameters change for roughing and finishing

*Roughing:* $c = 2,\ L = 50$ mm (includes overtravel), $D = 50$ mm

$$T = L + a = 50 + 2 = 52\text{ mm}$$

$$\text{N} = \frac{1000S}{\pi D} = \frac{1000 \times 25}{\pi \times 50} = \mathbf{159.15\ r.p.m.}\text{, say 156 r.p.m.}$$

$$t_r = \frac{cT}{fN} = \frac{2 \times 52}{0.25 \times 156} = \mathbf{2.67\ mins}$$

*Finishing:* $c = 1.\ L = 50$ mm, $D = 42$ mm

$$T = L + a = 50 + 2 = 52\text{ mm}$$

$$\text{N} = \frac{1000S}{\pi D} = \frac{1000 \times 25}{\pi \times 42} = \mathbf{189.47\ r.p.m.}\text{, say 156 r.p.m.}$$

$$t_f = \frac{cT}{fN} = \frac{1 \times 52}{0.25 \times 156} = \mathbf{1.33\ mins}$$

$t = t_r + t_f = 2.67 + 1.33 =$ **4 minutes**

N.B. Actual operation time will include set-up times, tool changing, non-machining feeds, etc. Cross reference the times on all planning sheets as shown for Operation 20, (a) and (b).

### *Example B*

Prepare a manufacturing package for the clamp shown in Figure 4.8.
Raw material 25 mm square mild steel bar cut into billets 55 mm long.

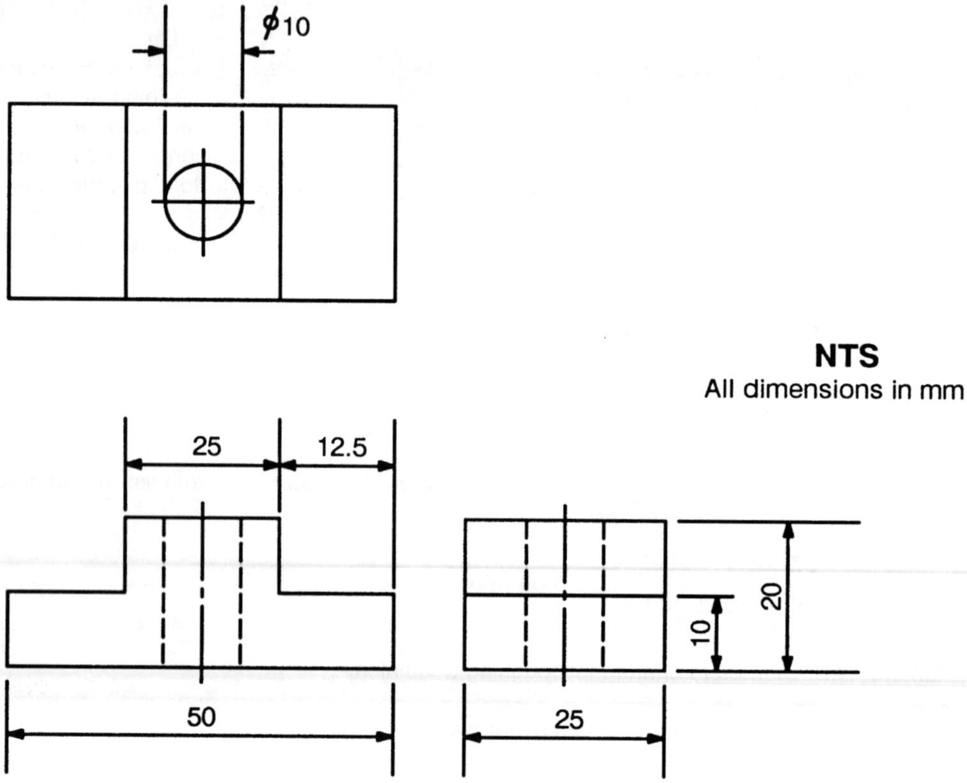


**Figure 4.8** *Clamp for Example B*

## ENGINEERING APPLICATIONS

## MANUFACTURING PLANNING PACKAGE

Part Number— EXAMPLE B

Revision— 1

Part Description— CLAMP

Material Description— 25mm x 25mm MILD STEEL BAR

Material Quantity— 55mm LONG BILLET = 1 COMPONENT

**Planner Details**

Name— J. RITCHIE　　Date 15 JUNE

Class— MECH. ENG. 1　　Project— 2

Page 1 of 2

## ENGINEERING APPLICATIONS – Planning Sheet

| Operation Number | Machine Description | Operation Description | Tooling Details/ CNC Reference |
|---|---|---|---|
| 10 | POWER | SAW MATERIAL TO LENGTH | MACHINE VICE |
| | HACKSAW | BILLET LENGTH 55mm | |
| 20 | VERTICAL | LOCATE IN MACHINE VICE | MACHINE VICE |
| | MILL | | |
| | | MILL RIGHT HAND END | Ø20mm SIDE |
| | | REMOVE 2mm | & FACE MILL |
| | | MILL TOP SURFACE TO HEIGHT | |
| | | COMPONENT HEIGHT 20mm | |
| 30 | VERTICAL | MILL OPPOSITE END TO | MACHINE VICE |
| | MILL | OVERALL LENGTH | Ø20mm SIDE |
| | | LENGTH 50mm | & FACE MILL |
| | | MILL STEPS TO SIZE | Ø15mm END |
| | | WIDTH 12.5mm | MILL |
| | | HEIGHT 10mm | |
| 40 | FITTING | DEBURR & EXAMINE | FILE |
| | BENCH | | |
| 50 | PILLAR | LOCATE IN MACHINE VICE | MACHINE VICE |
| | DRILLING | | |
| | MACHINE | CENTRE DRILL HOLE POSITION | CENTRE DRILL |
| | | DRILL THROUGH HOLE | 10mm TWIST |
| | | HOLE DIAMETER 10mm | DRILL |
| 60 | FITTING | DEBURR & FINAL EXAMINE | FILE |
| | BENCH | | |

Part Number – EXAMPLE B | Page 2 of 2

## 4.4 Process planning tutorial

*Exercise A*

1 Prepare a manufacturing package for the centre punch shown in Figure 4.9.
2 Calculate the machining time for turning the 15 mm diameter left-hand end.

Raw material 25 mm diameter high carbon steel cut into 130 mm lengths.

*Assume for all machining calculations that the lathe speeds available are 46, 64, 100, 156, 289, 409, 639, 1000 revs/min.*

| | |
|---|---|
| Feed = 0.25 mm/rev | Cutting speed = 25 m/min |
| Depth of roughing cut 2 mm max. | Depth of finishing cut 1 mm |
| | Approach distance = 2 mm |

*Hints*: running centre in tailstock can be used to support work and a split bush used to protect the knurl and locate work in chuck for finish turning.

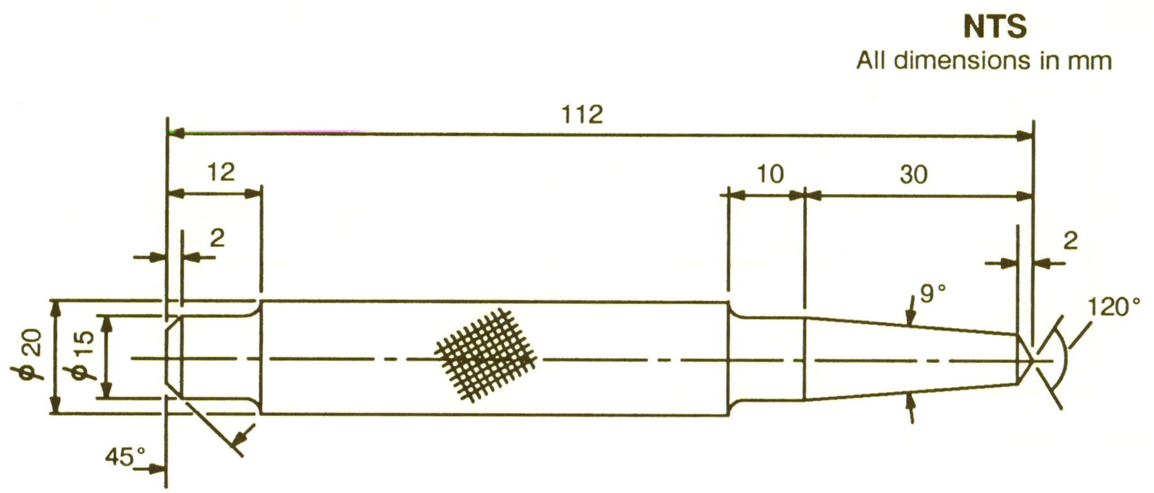


**Figure 4.9** *Centre punch for Exercise A*

*Exercise B*

Prepare a manufacturing plan for the support block shown in Figure 4.10.
Raw material 50 mm square mild steel bar cut into billets 145 mm long.

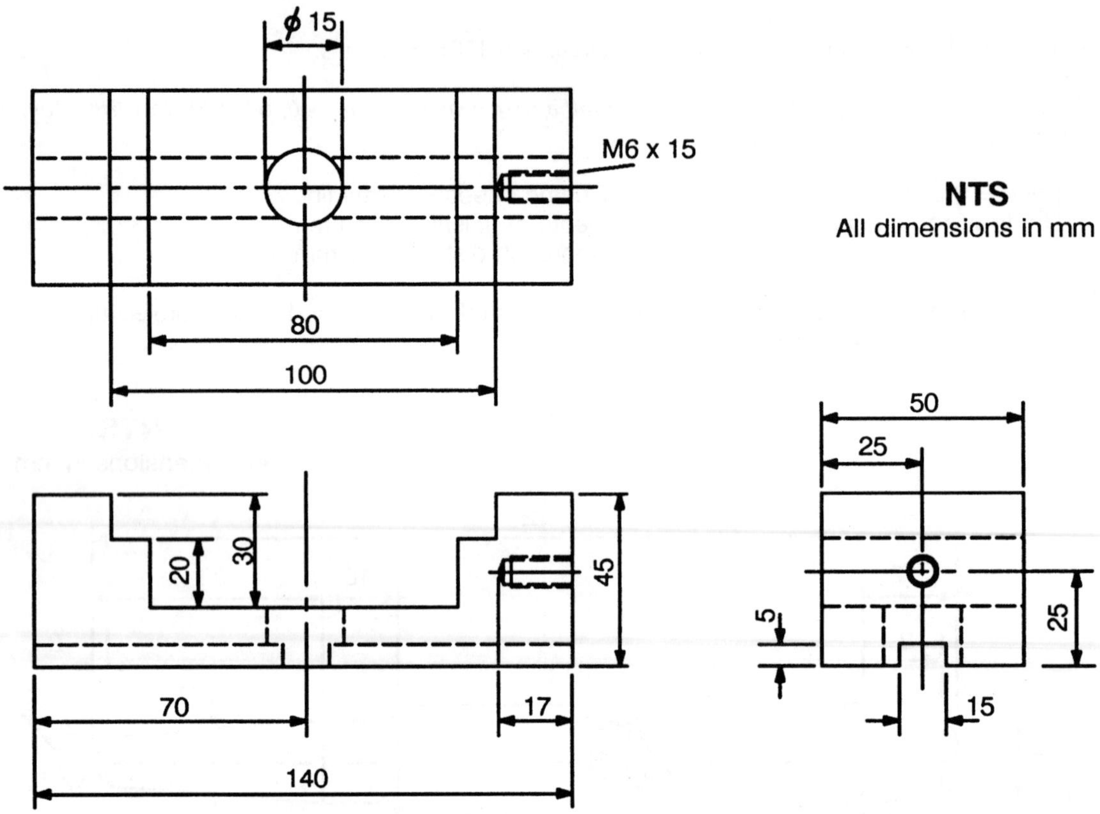


**Figure 4.10** *Support block for Exercise B*

*Exercise C*

For the leadscrew illustrated in Figure 4.11:

1 prepare a manufacturing process planning package
2 calculate the machining times required for the initial facing operation and turning the external diameter.

Raw material 35 mm diameter mild steel to be cut in 165 mm lengths.

*Assume for all machining calculations that the lathe speeds available are 46, 64, 100, 156, 289, 409, 639, 1000 revs/min.*

Feed = 0.25 mm/rev
Depth of roughing cut 2 mm max.

Cutting speed = 25 m/min
Depth of finishing cut 1 mm
Approach distance = 2 mm

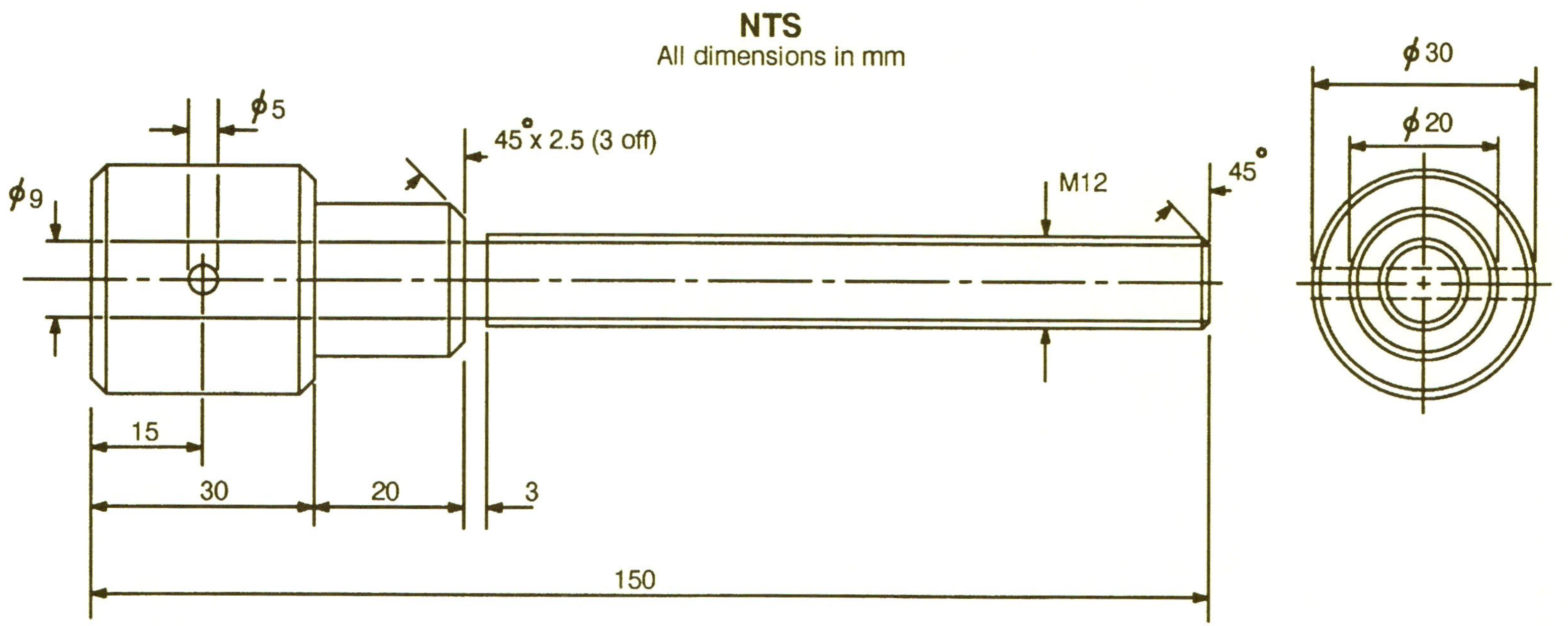


**Figure 4.11** *Leadscrew for Exercise C*

## References

1 *Introduction to Jig and Tools Design*, M. H. A. Kempster, The English Universities Press Ltd (1974, 3rd edition)
2 *Computerised Manufacturing Process Planning Systems* Hong-Chao Zhang and Leo Alting, Chapman and Hall (1994), pp. 1–44
3 *Manufacturing Processes*, B. H. Amstead, P. F. Ostwald and M. L. Begeman (1987), pp. 663–6

# 5
# Project outlines

Engineering applications are integrated into engineering courses to ensure that the student has the opportunity to apply the principles of engineering science to the solution of practical problems using materials and methods of manufacture typical of those widely used in industry today. Engineering is, after all, immensely practical but must be based on sound foundations of theoretical knowledge: 'engineering applications' provides a balance between theory and practice and are designed to 'keep the student's feet on the ground' . . .

This text identifies ten types of project which together can give a comprehensive introduction to engineering as it is practised in industry. Generally speaking, each consecutive project is built upon those which precede it, with sufficient overlap to reinforce and advance lessons already learned, so there is a progression in terms of complexity and technology from one to the next. That having been said, new aspects are introduced in each project directly from theory and also individual students tend to have different starting points, determined by their own personal educational and work experiences, so it is not necessary to stick rigidly to the order in which the project types are numbered and presented. Each project outline therefore has a paragraph entitled 'Prior experience and given information' so that an appropriate sequence of projects can be selected to suit individual students as well as individual courses. Each of the ten types of project are described in this chapter and their educational objectives identified.

The actual project specification given to you to work to will follow a similar outline. As you advance through each type of project, you will find that each one is generally less explanatory (and less helpful!) but more specific as to what precisely is required of you. The project specifications are provided in the *Engineering Applications Tutor's Resource Pack* and will be provided by your lecturer as hand-outs as you tackle each project type.

## 5.1 Type 1 projects An introduction to engineered projects

### *Outline and rationale*

'Type 1' projects are designed to introduce you, as a new student of engineering, to your new environment. The focus is on an engineering assembly, which is typical of your chosen branch of engineering. You and two or three of your colleagues will be required to take it apart to find out how it works and what it is made of. Each of you then has to select a part or aspect of the whole for private study, the results of which you will present in both oral and written form. As you follow through what is required of you by the detailed project specification, however, you will find that you also get to know the others you are working with in your group. More than that, you will start to establish a rapport with your academic tutor and some of the technical staff, as you find your way around the workshop or laboratory you are working in and learn to use some of the equipment.

It is assumed that you have little or no prior experience of the hardware and basic tools used within your particular discipline of engineering so, confronted with the initial task of disassembling an engineering artefact which you have probably never considered in detail before, you will soon discover the benefits of pooling the team's ideas and suggestions. Once you have decided how to proceed (a planning task in itself) and what tools to draw from the store, you will soon be handling components which are typical of the fundamental building blocks of a wide range of other assemblies, some of which may be designed for similar tasks but others may use the same technology for widely varying applications. So the hands-on experience you gain will stand you in good stead as a background to the understanding of the more theoretical aspects of your studies throughout your course.

As disassembly proceeds, you should try to bear in mind that your group is responsible for putting the whole thing together again in its original functioning state! So it is important to be methodical and observant as you lay out the individual bits and pieces. This will also help you to discover how the thing works and what part the individual parts play in its functioning so that when strip-down is complete, your curiosity will already be aroused and the choice of function or component for your own individual study will be easily made, subject to the agreement of the others in your group and of your supervisor. Remember when you are making this choice you can either look in depth at an individual component or sub-assembly, or look at the overall system. For example, you might have the opportunity of studying the impeller of a centrifugal pump which circulates water through a cooling system or you might study the cooling system itself: on the one hand you would become involved in considering the material chosen for the impeller, the casting process used to make it and so on, while, on the other, you would have to find out how the unwanted heat was picked up by the flow of water and then disposed of by some kind of radiator.

As you discuss with your colleagues which of you is going to study what, it is worth remembering that your individual study is part of the whole investigation so, when you come to make your oral and written contribution, it is a good reflection on your group effort if the sum of each individual's work adds up to a comprehensive picture of the functioning of the whole device.

**Figure 5.1** *A partially stripped motor cycle gearbox*

One final thought, although it is assumed that your group has virtually no prior experience of the engineering hardware in general or your allotted artefact in particular, and although it is probably the case that the assembly you work on is quite complex, nevertheless, a well designed product should be relatively easy to take apart and put together again. So, making due allowance for your collective lack of experience, you may well be able to give an informed opinion as to how well this aspect of design has been achieved in your particular product; it is surprising, and disappointing, how little regard many designers give to the maintenance aspects of their handiwork.

As far as your oral and written reports are concerned, you may consider yourself lucky if you manage to acquire a technical reference manual for the assembly you are working on but this can sometimes be a mixed blessing because such manuals often give far too much information which you will find difficult to reduce for a 6 or 7 minute presentation and a 1000 word report. Equally likely, they give lots of information about every aspect of the unit other than the one you are researching. For example, Figure 5.1 is a photograph taken of a motor cycle gearbox as a familiarization exercise proceeds. The workshop manual for it gives instructions about the dismantling and reassemble sequences and the tools required; it also details procedures for replacing seals and bearings and so on. But students working on it might well be struck by the fact that not only are the main casings made from aluminium alloy castings but many of the other components, such as the gear shifters and some of the gear clusters themselves, are also machined from castings made of other types of metals. Suppose one student decided to follow up on the casting processes and another on the materials chosen for the various components, each with its different functional requirements, then both would find that the workshop manual was completely unhelpful. In such cases it is actually easier to consult a general design textbook in the library and read up on the fundamental items such as bearings, seals, gears, clutches, lubrication, etc. which appear in the unit and then work out why the particular types and sizes found were chosen for the particular applications. Similarly, when researching the manufacturing processes used, it is almost always best to look at general manufacturing textbooks rather than try to find particular references to the particular component you are studying. Taking our earlier example of the pump impeller again, your are unlikely to find a textbook dealing with the manufacture of impellers but there are numerous texts covering the various casting processes which should enable you to determine how the impeller was made. Likewise all the other manufacturing processes, such as machining, welding, injection moulding, electrical discharge machining and so on, are extensively covered by books available in the library.

### *Time required*

Although none of the activities involved in this familiarization exercise are in themselves lengthy, they do all add up to perhaps something like ten hours of solid work. Given that you have deadlines specified and that there are marks penalties for late submissions, you would be well advised to schedule your time carefully. The following breakdown will help you:

- Strip down — 2 hours
- Investigate chosen area — 2 hours
- Rebuild — 2 hours
- Prepare presentations — 2 hours
- Attend oral presentations — 2 hours

Most of these are timetabled, group, activities so the time structure imposed on your schedule should make it easy enough to meet the deadlines provided you exercise a degree of self discipline in your own investigations and preparations for your oral and written reports.

### *Prior experience and given information*

No prior knowledge or experience is assumed.

### *Submitted and assessed material*

(a) Oral presentation. (Typically 7 minutes presentation per student with 3 minutes for questions: thus 12 students will make their presentations in a two-hour session)
(b) Written report. (Typically 1000 words plus figures and sketches)

### Additional information

For all workshop activities you must use personal protective equipment and your supervisor should insist on you doing so as this is now required by law. For this, and most of the projects described in these outlines, appropriate equipment will be overalls, safety shoes or boots with steel toe caps, and safety spectacles. Not only do overalls keep your clothes clean but, more importantly, they are designed to pull together loose items of clothing which otherwise might be caught in machinery: safety footwear saves feet when heavy or sharp items fall off the bench: safety specs shield the eyes from flying material, for example, high velocity particles coming your way when your inexperienced colleague is wielding a heavy hammer on a stubborn shaft, etc. In addition, if your hair is long, you should wear a hairnet especially if you are working with or near powered machinery. Caps with sown-in hairnets which are quite unobtrusive, can be bought and are widely used throughout industry.

## 5.2 Type 2 projects The design and manufacture of a simple part or assembly requiring some form of heat treatment

### Outline and rationale

'A picture is worth a thousand words.' Although it is true that most engineering designers would be unhappy to have their drawings referred to as pictures, the truth conveyed by this quotation is no where better exemplified than in the field of engineering endeavour. A vast amount of information can be conveyed with great accuracy by an engineering drawing, so drawings, whether produced by pencil on paper or by computer on screen, are the primary means of communication in every engineering discipline. But, as with a foreign language using a different script, the would-be engineer has first to learn the meaning of the individual hieroglyphics before being able to read and write fluently or, should we say, read and draw drawings. Engineering drawings have their own conventions as to how and where the various views of the object are laid out as well as conventional 'characters' which represent individual items or features. When a mechanical component, such as a shaft and bearing, is drawn, its representation is usually easily recognizable because it is the same shape and size (or scale of size) as the actual component but when a schematic diagram of an electrical or hydraulic circuit is drawn the symbols, or 'characters', bear virtually no relation to the shape or size of the units they represent. Fortunately the analogy with learning a language extends to the fact that, once a few basics have been understood, it is possible to start communicating and then pick up on 'vocabulary' and 'grammar' as you go. Also, fortunately for some of us, the analogy breaks down in that you do not have to be 'good at languages' to be able to learn to communicate in depth by means of drawings.

This type of project, then, is devised to introduce you to the application of the fundamentals of drawing including orthogonal projection, line type, dimensioning and drawing standard conventions, as well as the basic drawing equipment you will need to use for the more advanced projects and design assignments as your course progresses. Experience in all these aspects is gained as you design, draw and dimension a simple artefact such as a centre punch or screw driver as Figure 5.2. The detail piece-part drawing you produce will be the complete specification of the article using a minimum of words and numbers and, in fact, very few lines and circles. You will get the most out of this project if you subsequently have the opportunity to make and test the part. But more than that, this project is also designed to introduce you to the ergonomics of design, the metallurgy of heat treatment, process planning and manufacturing technology. Emphasis is on the word 'introduce' because, as you follow the four key stages of engineering a product, namely design, plan for manufacture, manufacture and deliver, in a short space of time, each can only be considered at an elementary level. None the less, if you gain some understanding of how they all dovetail together, you will find that knowledge to be an invaluable foundation for more advanced work in the next few weeks.

As far as design is concerned, this project demands little or no calculations; rather the design is confined to the ergonomics of handling and using the artefact. So if, for example, you were designing a centre punch, you would make it big enough for the user to get a good grip of it (and here the knurling helps) in such a way as to minimize the danger of the hammer striking the hand in use, but not so big as to be clumsy. The long taper section reduces the obstruction to the user's line of sight when the tool is being positioned so a relatively small angle is best. Finally the angle of the point is chosen to match that of the drill which will follow the punching operation. Nor is that all that 'Type 2' projects convey because the material from which the artefact is to be made, and, most importantly, its metallurgical performance, has to be considered. Continuing with the example of the punch, the point of the punch has to be hard enough to penetrate the

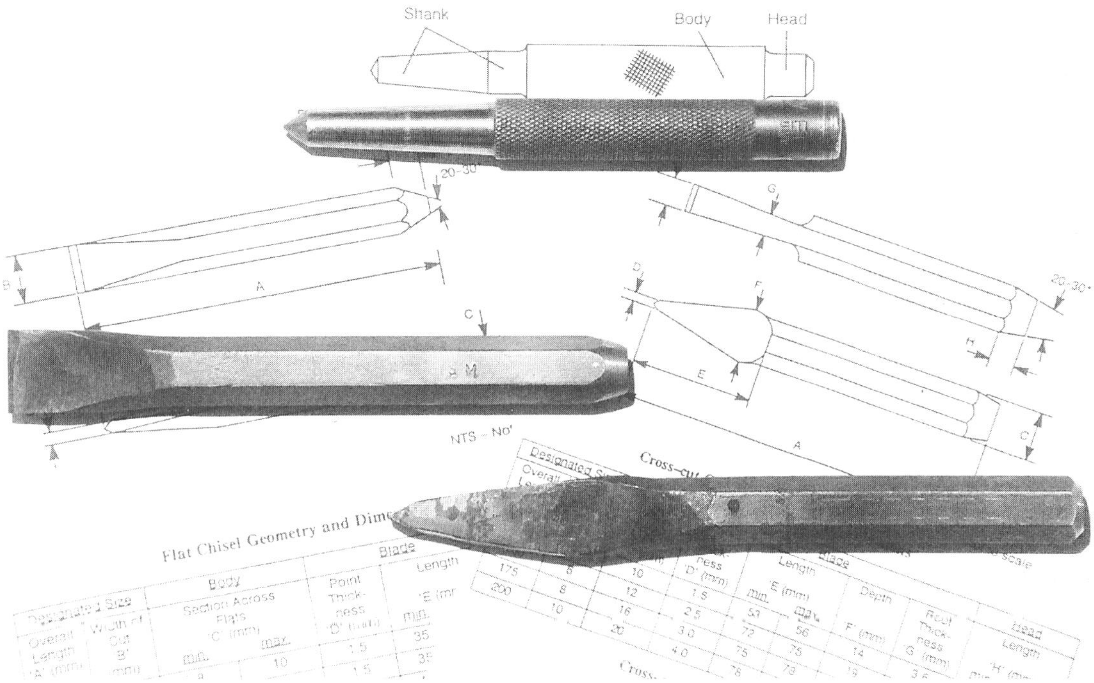


**Figure 5.2** *A centre punch or a chisel could be the focus of a 'Type 2' Project*

material being prepared for drilling and that material might itself be steel. It is an unfortunate fact of metallurgy that hard materials are invariably more brittle and less shock resistant than most softer materials. So the requirements for the business end of the punch are quite exacting: it must be harder than the material it is required to punch but not so hard as to be so brittle that it will shatter when struck. Then the other end of the punch has to be thought of because it is to bear the impact of the hammer. Again the specification works out to be quite exacting because this time it must be hard enough not to deform when struck but not so hard that particles will break off under the hammer blow. This last point is very important because particles which are knocked off by the hammer can travel at very high velocity and, although not big, can readily penetrate the back of the user's hand or, much worse, his eye, if he has been foolish enough to work without his safety spectacles.

Metallurgically then, the punch has conflicting requirements because the struck end must be somewhat softer than the opposite end. This conflict is resolved by making the punch out of a particular grade of steel or *alloy steel* which can be *hardened* by *heat treatment.* After machining to the sizes specified on the drawing, the whole punch is hardened and then the point, the struck-end and the main body are *tempered* back by differing amounts. (You can read up on the terms shown in *italics* in the materials guide to steels in the Quick Reference Guide.)

The experience of carrying a project through from concept to completion emphasizes the need for producing an unambiguous means of fully specifying sizes, surface finish, material and corresponding material properties (condition) to satisfy the ergonomic and performance requirements of the product. Consideration of manufacturing requirements leads naturally to process planning and, where appropriate, the need for hand processes and heat treatment during the development of the production sequence.

### Time required

- Drawing office work — 9 hours
- Process planning — 2 hours
- Workshop manufacture — 4 hours

Although most of the time indicated above will be dictated by your timetable, the specification defines three or four target dates which must be met so it is important that you keep checking your progress from stage to stage.

### Prior experience and given information

This project assumes no prior knowledge or experience. You should therefore receive instruction on drawing practice, process planning, forging, machining processes, heat-treatment processes and some hands-on workshop instruction as the project runs.

### Submitted and assessed material

(a) A rough outline sketch of the proposed design (this to be marked and signed off after discussion with your class supervisor)
(b) A final detail drawing of the item
(c) A process plan or set of plans for the manufacture of your product or assembly
(d) After testing, a short report on its functionality
(e) The finished product for assessment by your supervisor.

## 5.3 Type 3 projects The design and manufacture of a simple mechanical assembly

### Outline and rationale

This project involves the design and manufacture of an engineering assembly of perhaps six to nine components including moving parts taking you through all stages of the 'engineering product' cycle. Typically you will be required to design something like a machine vice or a toolmaker's clamp. The project will introduce you to assembly drawing, limits and fits, and screw threads and other fasteners whilst enlarging the scope for different manufacturing tools, machine tools, and other techniques. Again you will be going through the whole cycle of engineering a product and you should bear in mind that this is followed regardless of how simple or complex the end product is to be, whether a toolmaker's clamp or an aeroplane. Although you and your colleagues will all be working to the same specification, it has been arranged so that the required performance criteria are given in bands, thus permitting a diversity of solutions, allowing you to design your own unique product – a first step towards creativity and originality, and also towards making decisions when everything is not as cut and dried as we might imagine it should be. In fact this is the main difference between the more analytical subjects and design and manufacture; most of the practice of engineering follows the design and manufacture model because in real life situations we seldom are in possession of all the facts to the degree of precision we would like and there are almost invariably clashes between the required performance criteria which make compromises inevitable.

Other than cutlery and crockery, very few items of everyday household equipment are made in one piece. Even the simplest equipment, such as a can opener or a spade, is made as an assembly of several parts. In the case of a spade, the parts are fixed together permanently and are not required to move relative to each other; while, depending on the sophistication of its design, some of the parts of a can opener may well be required to articulate with the others. More complex goods, like washing machines and video recorders, obviously comprise a large number of individual components, each with their own function, some of which are fixed in permanent relationships with their neighbours and others which are termed 'moving parts'. The examples cited above are chosen because they are familiar samples of the very many engineered devices which we use from day to day. Across the wider spectrum of industry there is a vast array of equipment varying enormously in its degree of sophistication and complexity. When complex equipment is being designed, the designer will usually have to make use of a 'layout' drawing so that he can place each component and groupings of components in relation to the rest so that they will work together without getting in each other's way. Once the layout of the design has been established, the designer will create an assembly drawing or perhaps a number of drawings of sub-assemblies which show how the parts fit together and at the same time allow him to determine the final shape and size of each part. These drawings can be used later by the workers in the fitting shop or on the assembly line and also by maintenance personnel during the service life of the equipment. Actually

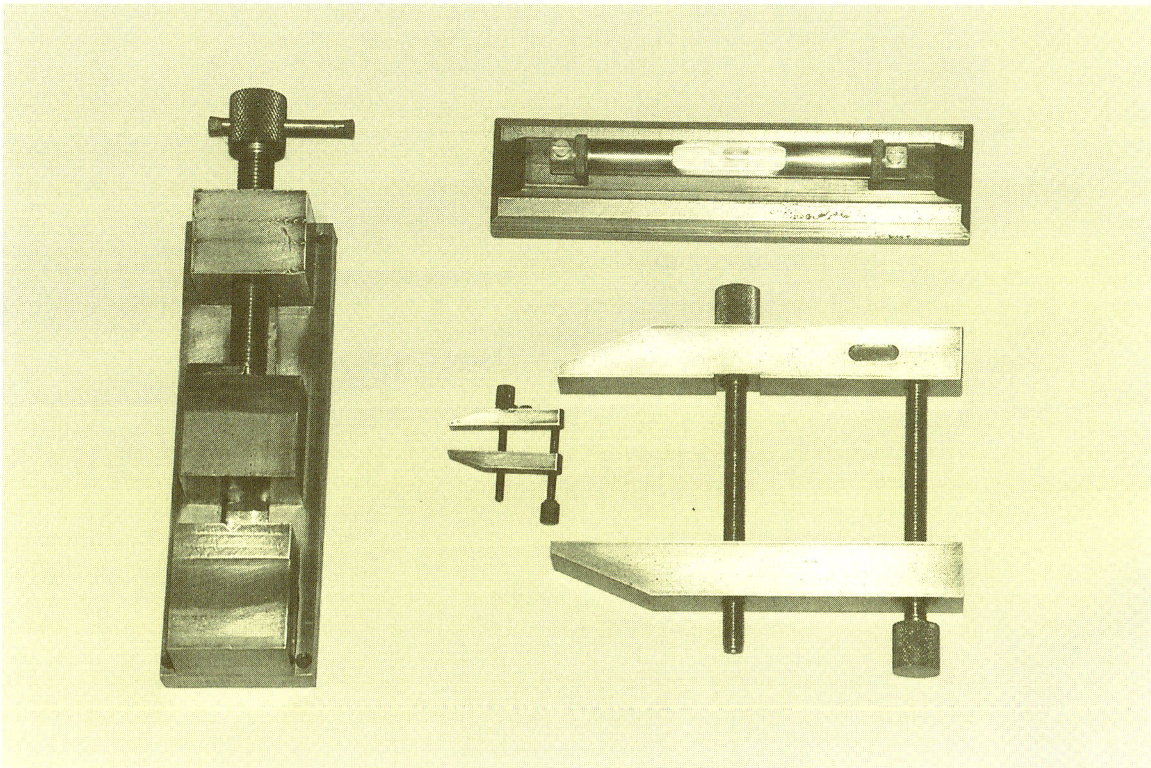

**Figure 5.3** *'Type 3' Projects involve assemblies with moving parts and screw threads*

things are often not that simple and the designer will usually have to change his layout drawing as he develops the sub-assemblies. Such changes frequently have knock-on effects which require modification to other parts and sub-assemblies, so the process may have to be repeated more than once as the design becomes more refined and finalized. It is from the assembly drawings that the detail drawings, or piece-part drawings, will be finally created to fully specify component parts.

The device described in the specification you will be given is not very complex (Figure 5.3 shows typical examples) so you will not need to produce a layout drawing before drawing the assembly. None the less, whether you are required to do so or not, you will be well advised to sketch the vice, or whatever, and try some of the key dimensions to find out how they interact with each other; the chances are that you will find yourself going around the design loop more than once. If you take this route, you should then be able to draw the assembly with little need to make changes as the design evolves. By the time you have finished the assembly drawing, you will have defined the overall dimensions of each component part so when you come to draw them as individual items you will not have to think about their size and so be able to concentrate on the details of how they will have to be specified to ensure that they will work together as required. This is where limits and fits come in.

Limits and fits are dealt with in some little detail in Chapter 10 of the Quick Reference Guide but for this project the requirements are straightforward and have been devised to give you an introduction to the concepts involved. You will find that each of your component parts only contacts others on one or two surfaces or at one or two places so the non-contacting faces are of little importance as far as the interaction of the parts with each other is concerned. These features, then, do not have to be made to tight tolerances since they do not 'fit' with others. Of the surfaces which do contact others, some are permanently fixed together and others have to move relative to each other. You will probably join the components which are to be fixed together with screws. The best way to do this is to drill clearance holes through one part and drill and tap corresponding holes in the other: this is tidier and uses fewer parts than the alternative of using bolts and nuts. You will find the required drill sizes given in the Quick Reference Guide Section 7.6 so the only other dimension you need to be concerned with for any pair of holes is the distance between them.

Since it is you who will be making the parts, you could ensure that the clearance holes through the first part line up with the tapped holes in the second by drilling the clearance holes first and then drilling the tapping holes in the second component through the clearance holes in the first. If you drill and tap one hole first and insert a screw to hold the parts together while you drill the second tapped hole, you should get a perfect result. In an industrial setting, however, where more than one, perhaps hundreds if not thousands, of the parts are to be made, this is not a practical method. The parts have to be made in batches, independently of each other, but when any mating parts come together they must fit: it follows that they are interchangeable, that is, any two components, selected at random, will fit together. To achieve this, the distance between the holes has to be carefully specified with a tolerance on the dimension which will ensure that any two parts will always fit when assembled together. That is why it is important that you grasp the concepts of limits and fits. Designing components within specified tolerance limits is at the core of all modern engineering production. It allows us to manufacture spare or replacement parts for products anywhere in the world so long as the manufacturing processes and equipment available is capable of achieving the accuracy specified for the part.

Generally speaking, parts which move relative to each other are constrained to do so in a particular path either by rotating around a shaft or pivot or by being guided in a straight line. Either way the accuracy of the movement required will determine the tolerances you choose. You will find by reference to Section 10.1 in the Quick Reference Guide that there are six recommended clearance/running fits. There is always the temptation to make the fit as accurate as possible to ensure the required performance but, as the final figure in that section shows, accuracy costs money. The designer should aim, therefore, to achieve the required performance with as wide a tolerance as possible.

Two other things which have to be considered when surfaces move in contact with each other are surface finish and lubrication. Obviously the smoother the surfaces, the smoother the movement between them and the less they will abrade each other: lubrication will have similar beneficial effects. These are two big subjects and are beyond the scope of this text. Again, however, it is worth referring to the two diagrams in Section 10.1 which show the surface roughness associated with a variety of manufacturing processes and which also make the point that there is a natural relationship between accuracy and surface finish, i.e. the tighter the tolerances achieved, the smoother the surface will be (but not the other way round).

You will not only be using screws to join parts together but also to make others move relative to each other. In this second context screws are seen as devices to change rotational motion into linear movement. More than that, this particular project focuses on the fact that screws can be used to apply significant forces. Your study of mechanics elsewhere will include the principle of the 'screw jack' which considers the mechanical advantage of the screw as a force amplifier. Screws are widely used in all kinds of industries to achieve these two results but they are not mechanically efficient because a lot of effort is lost in friction between the rubbing surfaces, so good surface finish is again important to get the best results.

Whether it is a vice or clamp that you are making, it is easy enough to arrange things so that the moving jaw is pushed closed by the driving screw. You will need to be much more ingenious, however, to make the screw open the jaws again. Circlips and roll pins are two useful allies here or perhaps you could utilize a spring; dimensions for a range of all three devices are given in Chapter 7 of the Quick Reference Guide.

Most of the components you design and make will be more demanding than anything you have done before so all through the process of designing and detailing you will have to bear in mind the manufacturing facilities which will be available to you and their capabilities in terms of accuracy and surface finish. This 'how to make' frame of mind is an important discipline to learn because you should employ it in all design work throughout your course and subsequent career. A study of 'Design details for machining' (Section 10.3) will pay dividends at this juncture and, if applied, will make things much easier for you when you come to make the bits. In any case the project specification requires you to plan the manufacture of at least some of the components you are to make so it is much more sensible to design out the difficulties as you draw than to face up to them at the planning or, worse, the manufacture stages.

### *Time required*

- Design and draughting 12–14 hours
- Process planning 5–6 hours
- Workshop manufacture 25–30 hours

The time required will vary depending on the number of parts in the product as well as on their complexity.

### *Prior experience and given information*

You should have successfully completed a 'Type 2' project or equivalent before embarking on this one.

You will receive general instructions regarding the interpretation of the functional specification, fastening methods, decision-making, recommended limits and fits, relevant manufacturing processes, etc. As indicated above, much of the detailed engineering information required will be found in the Quick Reference Guide. You should also have received manufacturing process information by the time you need to know about it. This should include information about machine tools with which you are not yet familiar, with a particular emphasis on precision manufacture to achieve the tight tolerances required for the various fits you will specify. By the time you come to make the parts, you should have received hands-on instruction in the use and reading of the micrometers and vernier calipers you will need to ensure that you have in fact achieved your own goals of precision.

### *Submitted and assessed material*

(a) A rough outline sketch of the proposed design and any calculations required by the spec. (These are to be marked and signed-off before serious work starts on the assembly drawing.)
(b) A finished assembly drawing of the product
(c) A set of detail part drawings, each component fully dimensioned on a separate piece of paper
(d) A set of three process plans including one of the assembly procedures for the product, one for a major turned part and one for a major milled part
(e) After testing, a short report on its functionality
(f) Your finished product for assessment by your supervisor.

## 5.4 Type 4 projects Finding out about manufacturing companies

### *Outline and rationale*

As we have stated earlier in this book, no economy, or the majority of people who live within its influence, will prosper without being able to design and manufacture products. It is a fact that the most successful economies in the world all have a very strong manufacturing base. Traditionally engineers have got on with engineering and left the management of their companies to others with different skills but it is now recognized that it is crucial that all engineers should have a good understanding of how manufacturing companies function and how they contribute to the economy of the country in which they are based. Moreover more engineers are required with the skills necessary to enable them to participate fully in the management of manufacturing companies. This project aims to give you an insight into how companies are run and financed; it particularly focuses on the financial accounting aspect of company management.

The understanding of company balance sheets, profit and loss accounts, and other financial measures will stand you in good stead as an engineer, whether or not you eventually move into the area of company management. It is important when key research, design and manufacture investment decisions are being made that you can follow the effects of these in terms of real company financial performance. To do so it is important to be able to understand the terminology used by the company's financial experts, otherwise it is easy to become confused and fail to grasp the facts. If its engineers are unable to contribute to the debate, there is a real danger that a company will be run by non-engineers who concentrate solely on the bottom-line financial figures.

This project requires you to examine an actual engineering company's published Annual Report and Accounts for a five-year period to determine how they have performed and what their performance trends are.

Another useful way of finding out about companies is to visit them and to learn how they design and make their products. Figure 5.4 shows a typical modern factory. In Chapter 1 we outlined the process of engineering a product and the manufacturing system. By visiting a company, you should be able to identify the various stages of the process and comment on them. If, in addition, you have had the chance to examine the company's financial records before your visit you will find it so much more enlightening. An early understanding of these concepts and of where they fit into a real environment is a very useful background for engineering students to have, with regard both to future work to be carried out during your course of

**Figure 5.4** *A modern 'high-tech' factory*

studies and also future employment. Companies tend to exhibit to visiting students their latest computerized methods of generating information and of controlling manufacturing machines and processes; your appreciation of the new technology in this type of environment will no doubt be enhanced thereby. But if, in addition, you have an overall view of the company's financial position your understanding of their position will be very much greater.

### *Time required*

| | |
|---|---|
| • Preparation of accounts or company visit organization | 2 hours |
| • Investigation of accounts/investigation of company | 2 hours |
| • Report writing | 2 hours |
| • Oral presentation preparation | 2 hours |
| • Own time | 2 hours |

Note that the times given here are based on *either* examining a company's accounts *or* visiting a company. If the two exercises are combined then at least 50 per cent more time will have to be spent on the project.

### *Prior experience and given information*

No prior experience is assumed for this project.

In preparation for the financial analysis project, you should be given a simple introduction to company accounting procedures, the balance sheet, profit and loss account and cash flow (source of Application of Funds). The key financial measures of which you should be aware are given in the project specification.

For the company visit you should understand fully Chapters 1 to 4 of this book. In particular, you should understand the information prepared at each stage of the engineering process and know about the types of products made by the company you are visiting.

### Submitted and assessed material

(a) Oral presentation (5 minutes)
(b) Written report.

## 5.5 Type 5 projects The design and manufacture of a load-carrying fabricated bracket

### Outline and rationale

This project will introduce you to welded and fabricated assemblies, welding techniques, design of fabrications, weld specification, simple structural analysis and mechanical testing. You will go through the 'engineering a product' cycle for a very different type of manufactured product than those involved in 'Type 2' and 'Type 3' projects but one which is very common in engineering practice, viz. the fabricated assembly. This can range from the small bracket designed as part of this exercise to the very large structures used in the off-shore industry.

The bracket, similar to that in Figure 5.5, has to be designed to carry a specified load at a particular position and in a particular direction relative to given fixing studs which attach it to its supporting structure: minimum weight of bracket is a mandatory design criterion. You will then fabricate the bracket to your drawing and test it to ensure that it is capable of carrying the design load. The test load may then be increased until the bracket fails so that its factor of safety can be estimated. The lightest bracket capable of carrying the design load is deemed to be the best.

Fabricated and welded assemblies differ from other types of assembly in that the various parts which are combined to make the whole are permanently joined together and cannot be sensibly taken apart again. Usually the parts are cut from stock material in the form of flat plates, bars (whether round, square, rectangular or etc. in section) tubular sections (again of various sections) or rolled sections (such as angle, 'Tee', channel, 'H'-section, etc.). Welding involves melting the surfaces to be joined whilst simultaneously

**Figure 5.5** *Fabricated brackets designed for a specific load-carrying duty*

filling the gap between the surfaces with molten metal of the same type as that of the parts to be joined; left to cool and solidify, the metallurgical structure of the welded joint becomes, theoretically at least, indistinguishable from that of the metal so joined. There are a number of different welding methods and techniques which you can learn about by reference to texts dealing with workshop technology. The method you will probably use is known as 'manual metal arc' (MMA). This involves striking an electric arc between an electrode, hand-held in a special tool, and the surfaces to be joined: the arc melts the surfaces and, at the same time, the business end of the electrode which thus becomes the filler metal.

Fundamentally the design of welds requires the size of the weld to be calculated so that it is sufficiently large to carry the load required of it without breaking or being stressed to a dangerous level. Theoretically a welded joint is as strong as the metal to either side of it but, in practice, there is always some doubt as to whether this is so because the intense heat may have weakened the metals or there may be unfilled gaps or other defects below the surface. Non-destructive testing of welds is expensive so the usual alternative strategy is to make the weld, say, three times stronger than it theoretically needs to be; this would give it 'a factor of safety' of three. To achieve this, sometimes the weld size is calculated and then made, in this example, three times bigger, but more often the figure for strength of the metal used in the calculation is divided by three and referred to as a 'design stress'. The end result is the same but, when using a design stress, the student may not realise that a factor of safety has automatically been incorporated in his calculation.

Design of fabrications can take this approach a bit further in that the good designer will first analyse his proposed structure to find the magnitude and distribution of the stresses imposed on it by the applied load and then strive to arrange the size and shape of the individual parts so that the joins are in the places where the stresses are lowest. By so doing, he will minimize the amount of welding required and generally end up with a minimum amount of material and hence a light structure.

Structural analysis in this project requires you to determine the magnitude and direction of the loads carried by each part and hence each welded joint: you can then determine the cross-section area of the parts and joints so that they are sufficiently large and strong enough to carry their loads. The direction of the load on each part is important because it determines whether the load, and hence the stress, is tensile, compressive, shear or bending. Calculations for the first three of these are more straightforward than for bending so it may be worthwhile trying to arrange the disposition of the parts to eliminate bending stresses on the welds.

Another point is that joints loaded in compression are least affected by internal defects in the weld, so they appear to be stronger. In industry this fact is usually neglected because the standard of welding is expected to be high since it is carried out by experienced welders. But for this project it is you who will be doing the welding and the experience of others shows that brackets which fail prematurely do so because of poor welds, so arranging parts so that the joints will be in compression, if at all possible, is a good approach.

There are two useful sections on welding in the Quick Reference Guide. Section 10.4 considers four aspects of design details for welding, while Section 8.3 gives basic definitions of welding terminology and then shows how your welds should be specified on drawings with symbols in accordance with national standards.

### *Time required*

- Design and draughting time 12 hours
- Process planning 4 hours
- Workshop manufacture 12 hours
- Own time 12 hours

### *Prior experience and given information*

Completion of a 'Type 2' project is an essential prerequisite. Also the fundamentals of mechanical assemblies should be understood before proceeding to fabricated assemblies so it is best to have done a 'Type 3' project as well.

Instruction will be given in sizing calculations. This will include structural analysis, reference to failure modes and welding practice with manual metal arc (MMA) techniques (fillet and butt welding), together with the use of welding symbols. This project should be reinforced with a parallel series of lectures touching on welding processes (particularly MMA) and the metallurgy of welding or it may be that you will have already studied these topics before reaching this project.

### Submitted and assessed material

(a) A rough outline sketch of the proposed design and your main calculations (these should be signed off after discussion with your supervisor
(b) A final assembly drawing of the bracket
(c) A set of detailed parts drawings, each part fully dimensioned on a separate piece of paper
(d) A process plan for welding the bracket
(e) The finished product for assessment by your supervisor
(f) After testing, a short report on its functionality.

## 5.6 Type 6 projects An air-mixing plenum: a sheet metal device incorporating a feedback device

### Outline and rationale

The device you will design mimics an 'economy cycle outside air' air-conditioning system for temperate climates. Warmed recycled air is mixed with fresh outside air in such a way as to maintain a required temperature of the combined flow. On warm days the flow of warm air is restricted by a manually set damper, while a motorized vane in the cold air duct is activated according to the temperature sensed on the down-stream side of the plenum so as to maintain the desired output. In hot climates, of course, air-conditioning plant is designed to handle refrigerated air.

This is a much larger project than any of the others considered so far and is designed to be tackled by a team of about three students. Most development work in industry is of such a size as to be too big to be tackled by individuals so, when working as a member of a team on this project, you will be in a situation very much akin to that of the average professional engineer. The main difference will be that there will be some dubiety as to the role of team leader. In a sense much of that role will be taken by your tutor but, unlike an industrial situation, he will probably leave you to work out amongst yourselves what each member's responsibilities are to be. Thus each member of the group will be responsible for a specific section of the work. That does not mean, however, that you are the only person to work in your specific area but rather that you will be responsible for that work being fairly distributed and carried out diligently and in time to meet due-dates. So in a real sense you each will be the team leader for some part of the project while at other times your role will be that of a team member. This experience of working in a group will impress on you the need for self-discipline, both when you are responsible for co-ordinating the work and setting targets and when you are managing your own workload so as to make a constructive and timely contribution to the development of the project and to the manufacture and testing stages.

Some buildings, vehicles, computer suites and other engineered environments require air-conditioning of one kind or another but, with the exception of vehicles and a few other standardized products required in large volumes, most applications are unique although they make use of similar units and devices to achieve the particular design specification. A company may specialize in developing solutions for such systems so, in terms of manufacturing products for markets, this project also demonstrates how engineering services and systems can be sold as tailor made.

The project introduces new ideas on three parallel fronts: the methods and procedures for drawing and geometry of surface developments and interpenetrations go hand in hand with sheet metal manufacturing techniques whilst maintaining the interest of mechanisms and moving parts in the active control required for the thermostatic control of the plenum. Your team will be required to design and manufacture not only the ducting which will merge and mix the two air-streams, together with the control vanes, but also the electrical/electronic circuit to drive the motorized vane. This control may be achieved either by an electronic circuit incorporating an operational amplifier or by writing a computer program to control the system directly via a personal computer (PC).

A vast array of products are made from sheet metal. The most familiar is undoubtedly the motor car but its pressed steel panel- and frame-work are not typical of the sheet metal work involved in this project because car parts are shaped in special purpose press-tools each made to the required geometry of each part. By way of contrast, the parts to be made for the ducting, vanes and other parts of the plenum (Figure 5.6) will be shaped for general purpose equipment. These are typical of products required in such low quantitites that the cost of special purpose tools cannot be justified.

**Figure 5.6** *Sheet-steel ducting manufactured with general purpose equipment*

You will be taught how to determine the shapes to which the thin metal sheet must be cut so that when it is bent-up will be of the required shape and size of the three-dimensional parts. Inevitably edges will have to be joined together by soldering to complete the part as well as to join it to its neighbours. Although not very strong in itself, the addition of solder to properly designed joints can make them extremely strong. These standard configurations for good joints are illustrated in Section 8.5 of the Quick Reference Guide; the allowances for the extra metal which must be added to the edges so the joints can be made, are also given.

The idea of thermostatic control is well known, certainly to everyone who lives in a centrally heated house. Usually with central heating the hot water or hot air is pumped around the system until the room monitored by the thermostat reaches a pre-set temperature level determined by the setting of the thermostat. At this temperature the thermostat sends a signal to the control which switches off the pump until such time as the temperature at the thermostat falls below a certain value, usually two or three degrees below the switch-off temperature. This is the essence of feedback control: a desired conditions is monitored and a signal sent to vary input parameters to maintain that condition.

Continuing with the central heating analogy, a second thermostat is incorporated in the system to monitor the temperature of the heating medium (the water or the air): it controls the heating source, turning the gas (or whatever) on and off in a similar manner. Obviously the heating medium has to be hotter than the temperature desired in the room monitored and the hotter it is, the quicker it will heat up the room. However the room heating system is often combined with a domestic hot water supply system and it is this which is often used to determine the temperature setting so that the risk of scalding is minimized. Therefore, it is this secondary factor which determines the speed of response of the system.

In this project, only one thermostat is used: it monitors the temperature of the mixed air delivered by the plenum. You will have to use the signal from it to automatically rotate a vane in the duct, so varying the quantity of cool air input and hence the temperature of the output air. The temperature of the warm air is constant but its rate of flow can be set by manually adjusting a vane in the duct which delivers it to the plenum. The amount of adjustment required of a vane to obtain a desired effect is not easy to predict because air is a compressible medium. This means that rotating a vane so that it obstructs half the cross section of a duct does not result in half the rate of flow; in fact the flow rate will hardly alter at all. It is unlikely that you will go into the theory of compressible flow at this stage of your studies but you will find, in working with your mixing plenum, that the motorized vane will be working much more towards the fully-closed end of its movement than towards the fully-open end.

You will have noted from this discussion, that, unlike a domestic central heating system, your system is to maintain the desired temperature by varying the rate of flow, not by turning air off and on. We have noted, however, that there is a 'dead band' of two or three degrees between switching off and switching on again. This dead band could be set to be much less but, if it were, then switching would take place much more often and so the control and pump would be subject to much more wear. This analogy can be applied to the control of your vane: if the control system were made too sensitive, the motor driving the vane would be constantly on the move, one way or the other. This is known as 'hunting' in feedback systems. In practice, then, a sensible 'dead band' has to be set as a compromise between accurate control and expenditure of energy and wear in achieving that control. As far as this exercise is concerned, however, a further factor comes into the equation, namely time. You will have to assess the effectiveness of your system in the laboratory over a range of temperature conditions. To do this, you will require a sensitive system if your test programme is not to take all afternoon.

### Time required

| | |
|---|---|
| ● Introductory lectures | 2 hours |
| ● Control theory and practice lectures | 4 hours |
| ● Mechanical design, calculations and sketching | 10 hours |
| ● Electrical/electronic design, calculations and sketching or computer programming | 8 hours |
| ● Workshop manufacture | 30 hours |
| ● System/circuit build and test or system build and programming and test | 6 hours |

### Prior experience and given information

You should have successfully completed a 'Type 3' project or equivalent since you will be expected to be familiar with assembly drawings and limits and fits. You will receive instructions as to how to handle the geometric constructions involved in the design of developments, interpenetrations and transition pieces while a parallel series of lectures will show what allowances are required for bends, soldered overlaps and other joints which may be required. Lectures in elementary control theory and the subsequent practical sessions will give you the information and practice required to tackle the control of the motorized damper.

### Submitted and assessed material

(a) A time-scale project workplan
(b) Detailed sketches of manual and motorized vanes with associated calculations
(c) A sketch of the baffle section
(d) Calculations and schematic drawings/program for electronic or computer control
(e) Detail part and assembly drawings for the manufacture of the complete system
(f) A final (group) report commenting on system performance in relation to predictions.

## 5.7 Type 7 projects An introduction to CAD/CAM

### *Outline and rationale*

This simple short project has been devised to give you a general overview of the use of computers in design and manufacture. You will gain hands-on experience of computer-aided design and manufacture (CAD/CAM) as you design a simple machined product and follow it through to computer numerical control (CNC) manufacture on a CNC machining centre. In fact you will take two routes to manufacture, one with a fairly low level of computer input and a considerable proportion of manual input and the other at a significantly higher level of computer technology and relatively little manual input. The idea is that you should appreciate that there is more than one way to programme manufacture by CNC machines. The mainly-manual route does not require much capital investment but does involve concentrated manual effort while the all-CAD/CAM route is much quicker (once you have mastered the computer software) but requires a substantial investment in both hard- and soft-ware. Thus you are introduced to two of the possible approaches to CAD/CAM commonly used in industry and which you will also require for later projects on your course.

Some typical designs of paperweights made to this project specification are shown in Figure 5.7. You will see that the 'design' aspects owe more to art than to engineering, the motifs being no more than slots machined out by cutting tools. You are required to create your own design and draw it to scale on graph paper. From there the low-tech' route involves choosing suitable cutters and a suitable sequence of operations and writing manufacturing instructions to the machining centre's computer in a code which it can then be used to run the machine automatically to cut out your component. The only use you make of a computer, other than that in the CNC controller, is when you type up your coded instructions. For the parallel route you must use a computer-aided design (CAD) package to draw your design onto the screen. You then use computer-aided manufacture (CAM) software to input your planned sequence of cutters and cutting operations. The software will then run a simulation of the cutting moves on the screen so that you can check all is in order.

**Figure 5.7** *Some typical designs of paperweights*

This project emphasizes how 'engineering a product' can be greatly enhanced by using computerized technology. The drawing of the part by CAD is the design function, the part programming of a machining sequence is the process planning function while the actual machining is the manufacturing function. As you take the CAD/CAM route you will become aware that the information required for manufacture is different from that required to create the drawing geometry in that the manufacture requires only the centre-lines of the slots, their widths being determined by the sizes of the cutters used. In other words, for manufacture, you need to program the cutter paths rather than the component geometry. To instruct the machining centre, you also have to input machining data such as what tool to use and spindle speeds and feed rates. So although the output at this stage is known as cutter location data (CLdata) there is more to it than just a sequence of location co-ordinates. A final step is needed because not all CNC controllers are coded in the same way. This means that the CLdata has to be 'post-processed' to make the instructions intelligible to the particular controller in the particular CNC machine tool you will use. The post-processor is a software package which re-formats the data to suit the controller: every different type of controller requires its own post-processor. These differences emphasize the need for different formats of data depending on whether the design or manufacturing function is being carried out. Data dependency is also emphasized in the simulation scenario. The main advantage of using this technology is that data is input once and then re-used as much as possible thereafter. You will realize, after completing the exercise, that it is much easier and more accurate to produce the cutting sequence by re-using the design data as opposed to re-creating the data in the form of a manually generated part program. The on-screen simulation also serves to boost confidence that no expensive mistakes are about to be made when the machine cuts metal.

This project represents the first steps towards an appreciation of the whole CAD/CAM scene but it should be borne in mind that there is more to this approach to engineering products than computer technology. Computers are merely the tools, albeit very powerful ones, which engineers have at their disposal. Their uses and limitations will become even more apparent as you tackle 'Type 8' and 'Type 9' projects both of which types are considerably bigger and wider than this introduction. This project has to be kept as simple as possible because, although we are looking at only a selected sub-set of the whole, for many of you this will be your first encounter with a bewildering array on new ideas and equipment. You are introduced to numerical control (NC) principles (as well as to the CNC machine tool available for machining), manual programming, the use of a CAD/CAM terminal for program preparation, the conversion of CAD drawing information into cutter location data (CLdata), post-processing and control tapes or distributed/direct numerical control (DNC).

### *Time required*

- Class time — 8 hours
- Private time — 2 hours
- Workstation time — 3 hours
- Machine time — 1 hour

Note that having completed your own CAD/CAM workstation session (1 hour) you will be required to tutor a colleague undertaking the same exercise: that is when the learning really takes place.

### *Prior experience and given information*

To obtain the full benefit from this project, you should have satisfactorily completed a 'Type 3' project (or equivalent) including at least the planning and manufacturing aspects. It is important to understand the fundamental manufacturing concepts, e.g. milling, before proceeding to computerized applications.

At the beginning of the project, you will be faced with a confusing amount of new technology, terminology and handout material. Do not be put off by this. By the end of it you will have a thorough understanding of all of this and will be able to tackle more complex projects in the CAD/CAM area with confidence.

### *Submitted and assessed material*

(a) A manual sketch of your proposed design showing the key ordinates and the depths of the slots for its manufacture. This should be discussed with your co-ordinator
(b) A CAD component drawing of the item (not necessarily fully dimensioned)

(c) A CAD manufacturing drawing of the key geometry required for computer aided part programming
(d) A manually produced part program to produce the part
(e) A post-processed output for the part's manufacture produced using the CAM package
(f) A short report covering the key CAD/CAM concepts experienced
(g) The finished machined component for assessment by your supervisor.

## 5.8 Type 8 projects The design, manufacture and testing of a complex component using CAD/CAM

### *Outline and rationale*

'Type 8' projects are comprehensive design–make–test group projects in which the design interest focuses on the functioning of a critical component in an engineering system: the design, planning for manufacture and manufacturing emphasis is in the CAD/CAM area. A typical component might be the impeller for a centrifugal pump (see Figure 5.8) or a lift-pad for an air-bearing system. The performance of the component is subsequently tested in the laboratory in some detail, making use of a test rig specifically provided for the purpose. The time you will spend at the CAD/CAM workstation provides an opportunity for developing further your proficiency in taking a design from the creation of geometry through the planning process to the actual production of the part.

Because you will be concerned with a complex part which would be very difficult to make on manually controlled conventional machines, you will come to appreciate how computer numerical control takes much of the skill away from the machinist but requires other skills of the planner/programmer. Also, because planning and programming are usually done remotely from the machine tool, you will quickly realize that to carry out these functions efficiently you must have good knowledge and experience of machining in general and CNC machining in particular.

**Figure 5.8** *Centrifugal pump impellers made to various specifications*

The reasons for saying this are twofold. In the first place, the opportunities for the operator to intervene and make adjustments are limited, especially in an industrial situation where the operator is seldom the one who does the programming. When you are working on a conventional machine, you can try out cutting conditions such as feed, speed and depth of cut. You soon learn when you are being over-ambitious and should back-off or, alternatively, when things are going well, you can save time by stepping up one of the factors. This first hand 'feed-back' is lost when cutting conditions are programme remotely. Usually the feeds and speeds are embedded in the CAD/CAM database used; that is to say, when you select a particular tool, the software automatically provides the recommended feed and speed as soon as you instruct it as to the type of material you are to cut. All that is left for the programmer is to decide the depth of cut, usually in terms of the number of cuts to take. The inexperienced programmer tends to be over-cautious or quickly becomes so, once he has been responsible for a broken tool or wrecked job.

The second reason is that the CNC machining centre has much more power to its spindle than its conventional counterpart. To put this power to good use cutting speeds, feeds and depths of cut should be maximized; but when they are, the cutting forces are also markedly increased. So it is then important to ensure both that the cutting tools chosen are suitable for these more extreme conditions and that the work holding and fixturing is suitably rugged. It is very probable, then, that most of these practical aspects will be provided for you in the project specification.

In fact it is always good to recognize the importance of considering fully the manufacturing constraints of the equipment available before progressing with product design and development. You can then design for manufacture in such a way as to maximize the full potential of the manufacturing equipment whilst avoiding, as far as possible, its limitations. That way you will produce the most cost-effective designs.

It is quite likely that this approach will have to be adopted when you come to the CAM aspects of the job, too, because the software will have its own peculiarities and limitations. For example, most CAM software for milling will offer an area-clearance routine which will allow you to define a boundary and then to leave the software to generate a series of cutting movements to machine away all the material within that boundary to whatever depth you specify. Probably this routine will claim to cope with islands; that is to say, you can specify lesser boundaries within the original one and define them as islands; the software will then carry out the area clearance but leave the islands untouched. This is very useful if, say, you are designing an impeller with a number of blades: you could specify the blades as islands and the software would automatically machine around them for you. In practice, many software programmes are limited to taking into account only one island at a time so, typically, you might find the machine produces the first blade perfectly then destroys it when dealing with the second, and so on. If the package you are to use has this limitation, then your instructor will teach you a cunning dodge or two to work your way around it and so get a perfect result.

More positively, you should notice that the geometry created on the CAD system is used and redefined for the CAM stage. The importance of the re use of data is thus again underlined and will be fundamental to your knowledge as an engineer in any modern manufacturing company. Together with the recognition that you have once again completed the 'engineering a product' cycle, this will be one of the most important outcomes of this project.

The use your component is put to in the laboratory will depend largely on how well this project is integrated into other aspects of the course. Even if, or especially if, the test programme is limited, you should try to find out how well your product performs compared with those made by the other groups. It is important not to lose sight of the fact that the production of components is not an end in itself; they are made for a purpose and assessing performance in fulfilling purpose is fundamental to all development work.

### Time required

| | |
|---|---|
| ● Design and sketching | 6 hours |
| ● CAD/CAM | 4 hours |
| ● Manufacture | 1 hour |
| ● Testing | 1 hour |
| ● Analysis of data, calculations | 4 hours |
| ● Write report | 4 hours |

### Prior experience and given information

The basic requirements are CAD/CAM awareness of the kind involved in 'Type 7' projects, together with the drawing practice, process planning and manufacture gained in 'Type 3' projects. An elementary introduction to compressible and incompressible flow, followed by an outline of industrial practice will provide the background context. To this will be added sufficient theory and guidance to allow the required design estimations to be carried out and hence the parameters for the required performance to be determined.

You will find that the routine for testing and the readings (range and increments) to be recorded are defined in the test instructions, as are the calculations based on them and the method of presenting the results.

### *Submitted and assessed material*

(a) A fully dimensioned drawing of the component (CAD)
(b) A manufacturing drawing (cutter centre-lines, etc. for CAM)
(c) A computer animation of the machining sequence (to be viewed on screen by your supervisor)
(d) Test data and analysis
(c) A final group report with calculations, analysis, comments and recommendations.

## 5.9 Type 9 projects The role of detailed analysis in engineered products

### *Outline and rationale*

You will work on this project with one or two of your colleagues. Compared with earlier projects, the emphasis here now shifts to the design calculations involved in striving for an optimized frame structure to carry a specified load. As an example, a frame structure, designed to carry 4 kN, is shown in Figure 5.9(a), being tested in a tensile testing machine. What is not clear from the photograph is that the frame is designed to carry a load from one point in a structure to another around an obstacle; that is to say, the load line is interrupted by another member so a frame has had to be developed to carry the load. Obviously, if it were not for the obstacle, the load could have been transmitted by a straight link.

Frame structures are not typically found in engineered consumer products but they are very common in industrial engineering practice although usually somewhat bigger than the one you will be designing. You can see that the frame shown has been made from lengths of angle material riveted together, using gusset plates at the junctions. The manufacturing aspects are therefore simple and unsophisticated, although there are one or two tricks of the trade and short-cuts to be learned as you join the bits together: these you will find as you follow through the manufacture and assembly instructions detailed in the project specification.

You will be instructed to treat the frame as a pin-jointed frame when doing your design calculations. Pin-jointed frame theory, as its name suggests, is based on the assumption that each member of the frame is joined to its neighbours at both ends by single pins which carry force only and no bending moment. In practice, for the frame you design, the members are not joined directly together but via gusset plates and you will find that you have to use more than one rivet ('pin') at each joint. Consequently they will transmit bending as well as force. The theory to allow for this is considerably more difficult than pin-jointed theory so the easier approach has been chosen but it has to be realized that your analysis and calculations will be at best a close approximation. The specification you will be given will encourage you to make the frame simple and light-weight, using as few rivets as possible, because the fewer the rivets in each joint, the closer they will be to 'pin-joints'. When you try out some rough calculations of possible configurations you will find that, in general, if you make the members relatively long and far apart, the forces carried by each are lower and so the number of rivets required are fewer and the gusset plates smaller. This strategy, aimed at keeping as close as possible to the pin-jointed ideal, has two drawbacks. One is that long members increase the weight of the structure: the other is that those members which carry compressive loads will buckle more readily. So compromises have to be made and the grounds on which they are to be made are not all that clear cut. No doubt you will have some interesting discussions with your colleagues: this is why no two groups will design identical frames.

There is a great deal to be learned from the testing of your structure and those designed by other groups; make sure you compare their performance with yours and try to understand why some do better than others. What you will discover is that, as the load applied during test increases, many of the frames will twist out

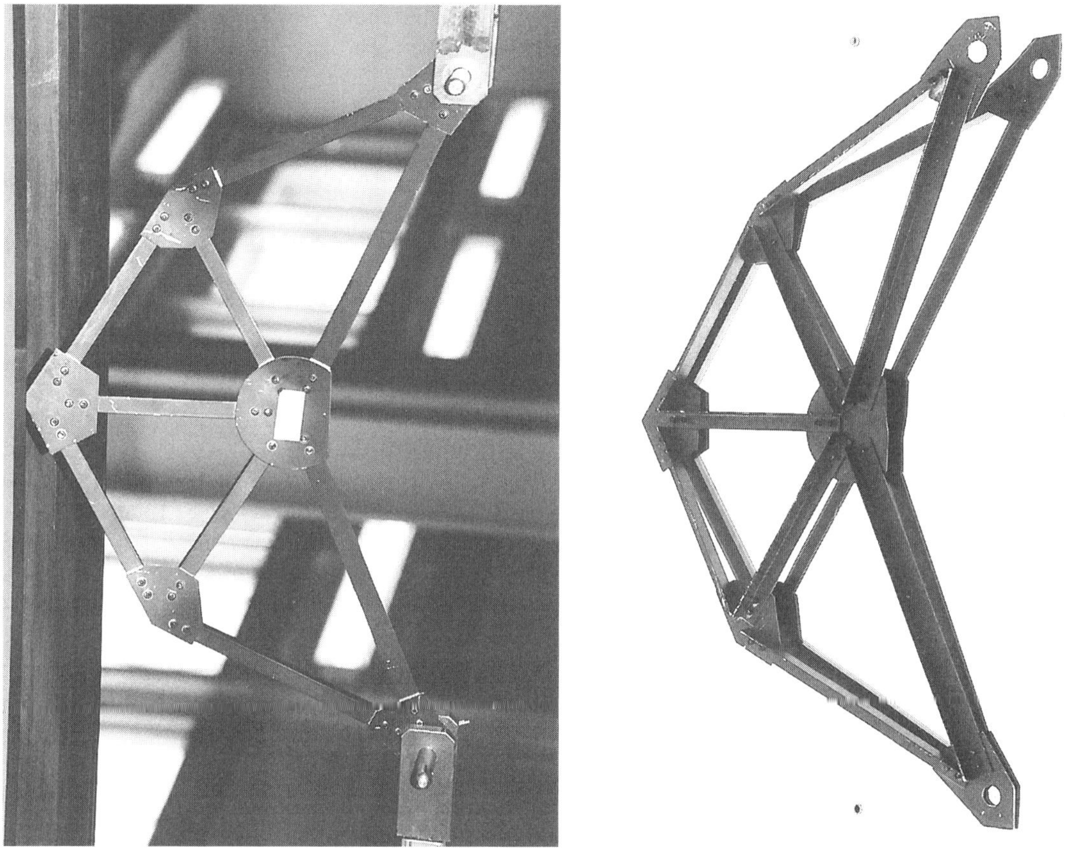


**Figure 5.9** *(a) A frame structure undergoing test in a tensile testing machine. (b) Laid upon a flat mirror surface, the reflection of the frame emphasizes the out-of-plane distortion which leads to premature failure*

of plane and fail prematurely. These early failures occur because the design calculations assume that the load is carried literally in the plane of the paper you have drawn the structure on. In other words the theory does not allow for the fact that the structure has thickness and, more importantly, that it is made out of angle rather than flat strips of material. Consequently the side view of your frame is not symmetrical and the load is effectively applied off-centre. You can reduce the likelihood of this twisting happening to your frame by adopting the back-to-back option with the angle. This will result in the load being applied on-centre, although the weight of the frame may increase somewhat despite the fact that doubling up each member means that you can use angle of smaller cross-sectional area. Manufacture and assembly will also be a little more difficult and time consuming.

### *Time required*

| | |
|---|---|
| ● Determine frame configuration | 16 hours |
| ● Detailed calculations and detailed frame assembly drawing | 12 hours |
| ● Detailed parts drawing | 8 hours |
| ● Testing and final report | 8 hours |

### *Prior experience and given information*

Although limits and fits do not feature in this project, you should be familiar with assembly drawings and auxiliary views as demonstrated in 'Type 3' projects. The frame is treated as a pin-structure: you will have to use either the method of joints or the method of sections for analysing determinant sections. The principal modes of failure of pinned joints will be explained in the supporting lectures/tutorial sessions. Manufacture and assembly utilizes little more than a hand-held drill and a *'pop'*™ *rivet* gun (details of *'pop'*™ *rivets* are given in Section 7.3 of the Quick Reference Guide). It is important, though, to use the procedures given in the specification to ensure that the frame all comes together, with each of the holes in the right place.

### Submitted and assessed material

(a) Sketches and selection calculations for at least three possible configurations
(b) Frame assembly drawing with detailed calculations for every member and gusset plate
(c) Detailed parts drawings
(d) Calculations of weight, material costs and labour
(e) A final (group) report.

## 5.10 Type 10 projects Consolidating the 'engineering a product' cycle

### Outline and rationale

The main purpose of this book has been to introduce you to the various stages and activities involved in the engineering of products so vital to the prosperity of manufacturing nations (see Figure 5.10). An outline of this overall concept, how it is carried out within a company and the departments, people and systems involved have been described in Chapters 1–4. All this has then been reiterated and reinforced as you have carried out the various projects. What is now required of you by this final project is that you consider two products, one from the Projects 1 to 9 and the other a commercially engineered product, and consider the role of the engineer in their development. The design projects presented in this book have enabled you to carry out and experience all the engineering functions within the 'Engineering a product' cycle from the early stages of design through to the delivery to the customer (who happens to have been you in each instance). The understanding of how this experience integrates with the information given in the earlier chapters of this book is vital for you in developing a secure foundation of knowledge and awareness of the

**Figure 5.10** *Exports are a measure of competitiveness in the international marketplace*

role of the engineer. 'Type 10' projects are specifically designed to underpin this understanding by making you think about this link as well as requiring you to explain it to, and discuss it with, your colleagues. Working in teams of two or three, you will prepare a group report and a joint presentation to sum up your Engineering Applications work. By the time you have reached this stage of the book, you will have completed a comprehensive group of design projects and so should be well prepared to tackle more difficult engineering tasks, always bearing in mind the lessons you have learnt from this early experience of engineering.

### Time required

| | |
|---|---|
| • Investigation of stage 1 | 2 hours |
| • Outline and discuss conclusions with your supervisor | 1 hour |
| • Report writing | 2 hours |
| • Investigation of stage 2 | 4 hours |
| • Outline and discuss conclusions with your supervisor | 1 hour |
| • Report writing | 2 hours |
| • Prepare presentation | 2 hours |
| • Give/attend group presentations and discussions | 2 hours |

### Prior experience and given information

You should have completed at least five of the design projects outlined in this section as well as fully revised and understood Chapters 1 to 4. You will be allocated a commercial product for your group's consideration by your supervisor.

### Submitted and assessed material

(a) Oral presentation (10 minutes plus 5 minutes for questions).
(b) Written group report.

# 6
# Writing reports

## 6.1 Introduction

Work, however well done, is of little value unless its results are effectively communicated to others, either in writing or orally. This is particularly the case in an engineering environment. During your course you will be asked to write technical reports (laboratory and otherwise) to develop your communication skills. You will probably find it difficult to write a good report early on and practice is required to achieve a decent standard.

A report is a formal, considered document. Its aim is to present information clearly and concisely so that it may be easily understood by the reader. For most of your reports you will have to assume that the reader has a technical background, but is not familiar with the subject matter; for others, the reader could be either completely familiar with the subject or possess minimal technical education or expertise. Your report should cater for the intended audience.

The suggested format relates to technical reports but can be amended to suit other types where necessary. The idea behind this discipline is to get you routinely to produce consistently high-quality, well-structured reports to a uniform standard.

## 6.2 Preparation and general hints

Work out a set of themes and then a rough contents list for the report and build up its layout around these. Make sure that the layout of the report is logical and that a theme is neither repeated nor split. The following order is suggested for the major headings of the report:

- Summary (maximum half of a page)
- List of Contents
- Introduction
- Objective (this may be combined with Introduction)

### *6.2.1 Body of report*

Contents are variable but should include (not necessarily in the indicated order) such headings as the following. For technical reports:

- Theory
- Apparatus
- Experimental Method
- Results

For non-technical reports:

- Background
- Methodology or Method of Investigation or Approach
- Analysis
- Results

Reports usually end with the following sections:

- Discussion
- Conclusions
- References
- Appendices

Discussion and conclusions should form roughly 25 per cent of the main body of the report.

Prepare a draft of the report based on this layout, read the draft a day or so after writing it and rewrite it as many times as necessary until you feel that you have produced a script which clearly communicates your work.

## 6.3 Guidelines

### *6.3.1 Layout*

Choose a *short* title for the report; use a subtitle if necessary. The major headings must be given titles and may be numbered starting at the Introduction. Subheadings may also be used if required, and a consistent numbering or lettering system may be adopted, for example:

1 Introduction
1.1 First sub heading
1.2 Second sub heading
1.2.1 Sub-sub heading

### *6.3.2 Style*

Try to write precisely. Do not construct sentences of excessive length, nor use sentences which are unnecessarily brief. Avoid vague expressions, unwarranted brevity and, in particular, colloquialisms. Do not use lists as an excuse for not constructing sentences. Reports should be written in the past tense and passive voice when describing apparatus or what happened, e.g. 'The engine drove the brake and its power was measured' not 'the engine drives the brake and I (or we) measured its power'. When describing relationships or results the present tense may be used, e.g. 'Figure 3 shows that engine efficiency rises with power'.

Remember that the report is *not* an instruction manual or a diary!

### *6.3.3 Presentation*

#### *6.3.3.1 Writing*

A report should be neatly hand-written, typed or word-processed. The total length of a report should not exceed about 10 pages or 2000 words excluding figures or tables. If in doubt about spelling, consult a dictionary!

*6.3.3.2 Diagrams*

The use of *simple* diagrams is encouraged where these are effective in conveying information concisely to the reader. Diagrams should be carefully drawn on plain paper and labelled, with important dimensions if appropriate. Each diagram should be given a figure number and title, and should be accompanied by a brief explanatory passage in the relevant section of the text. The front part of the number should relate to the numbered section in which it is placed. The final part of the number refers to its position in the section. For example **Figure 1.1** is the **first** figure in section **1**, **Figure 1.2** is the **second** figure in section **1**, and so on.

E.g.

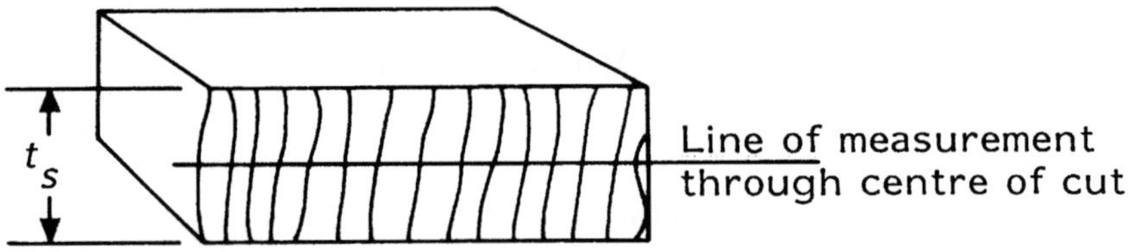


**Figure 1.1** *Determination of* $R_a$ *value for cut surface*

*6.3.3.3 Tables and graphs*

It is essential that results are clear and concise for the reader. Tables and graphs, if properly presented, enable large amounts of data and information to be conveyed effectively; they should be used instead of repetitive calculations to show important results. Readings taken from an experiment are often better presented on a graph.

Each table should be numbered and given a title, above or below. Each column of the table should be identified by the quantity being measured, units and a scaling factor where applicable; the same numbering system applies as for figure numbers.

**Table 1.1** *Load and extension values for test specimen*

| *Load (kN)* | *Extension (mm)* |
|---|---|
| 3.2 | 0.1 |
| 5.7 | 0.5 |
| 8.1 | 1.3 |
| 10.6 | 1.9 |
| 12.0 | 2.3 |
| 14.4 | 3.0 |

For graphs (see below), the following points should be noted:

(a) Care should be taken in selecting the axis scale, particularly noting whether the origin should be included.
(b) Label axes clearly with the quantity and the units.
(c) If the points on the graph are visible, distinguish between different sets using different symbols.
(d) Print on graphs as it is much clearer than script.
(e) Give the graph a figure number and title. It should be accompanied by a brief explanatory passage in the relevant section of the text.

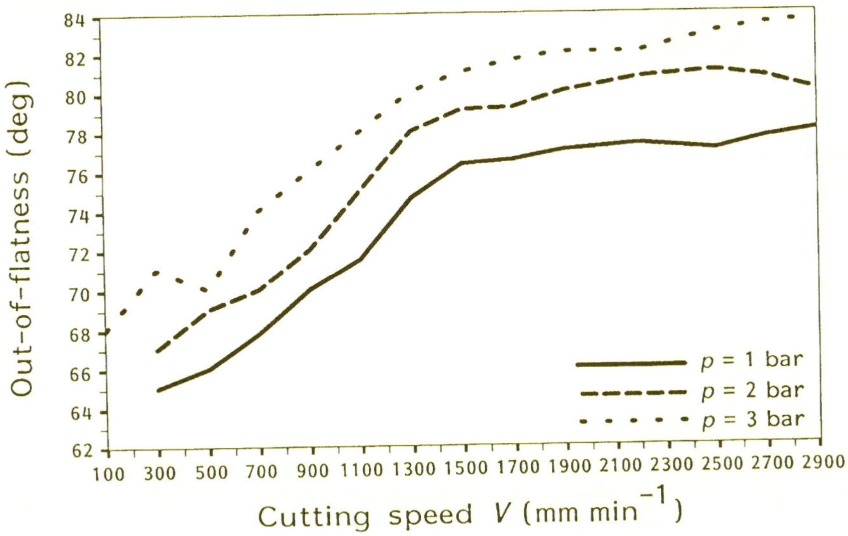


**Figure 1.1** *Determination of* $R_a$ *value for cut surface*

*6.3.3.4 Covers*

The report should be *firmly* bound with a standard cover. Allow sufficient margins for binding. The title, date and author's name should appear on the front cover.

## 6.4 Report contents

### *6.4.1 Summary (technical and non-technical)*

The summary (sometimes called an abstract or synopsis) must appear at the beginning of the report. It is the most important section, since it is used by a reader to decide whether he or she is interested in the rest of the report. Always write the summary last. It should be clear and specific (quoting numerical values if appropriate). It should be about 100 words in length.

### *6.4.2 List of contents (technical and non-technical)*

This should contain a list of headings, subheadings and appendices which appear in your report, starting with the introduction and the page numbers on which the heading appears.

### *6.4.3 Introduction (technical and non-technical)*

This should deal *briefly* with the historical or technological background to the report where applicable. It should describe the relevance and application of the work. The aims and objectives should be clearly stated.

### *6.4.4 Objective (technical and non-technical)*

This should state *briefly*, but clearly and concisely, the main reasons for the investigation; it may sometimes be included in the introduction.

### *6.4.5 Body of the report*

*6.4.5.1 Theory (usually technical)*

This section should cover only theory which is relevant to the experiment. It may be quite short for a laboratory report. It can sometimes be included as part of the introduction. Well-established theory can be quoted and does not require a reference. Less well established theory can be quoted and must be referenced, e.g. 'The theory developed by Smith[2] shows that . . .'

#### *6.4.5.2 Apparatus (usually technical)*

Do not use lists. Describe only more unusual apparatus and make use of simple diagrams as aids to their description.

#### *6.4.5.3 Experimental procedure or method (usually technical)*

Avoid use of lists of instructions as procedure. Do not describe experimental methods in excessive detail but mention essential or special techniques. Describe what was done and what happened.

#### *6.4.5.4 Background (usually non-technical)*

This may include the objective of the report. Background information relating to the scope of the subject being investigated, current methods/techniques/applications, refrence or case-study material may be included.

#### *6.4.5.5 Methodology or Method of Analysis or Approach (usually non-technical)*

This would include an outline of the way in which the problem or subject matter was investigated along with key assumptions, reference sources, etc.

#### *6.4.5.6 Analysis (usually non-technical)*

If statistics are involved this may take the heading Analysis of Results. If it is a discussion section then the key arguments developed as a consequence of the assumptions and methods used in the previous section would be outlined.

#### *6.4.5.7 Results and calculations (usually technical)*

All analytical results must be presented in SI units. Make sensible use of tables and graphs. If you wish to include large numbers of results, relegate these to the Appendices with a reference in the text. No simple or repetitive arithmetic or algebraic working should be shown.

#### *6.4.5.8 Discussion (technical and non-technical)*

The discussion forms the most critical part of your report and should reflect your individual interpretation of the work described in the report. It should be used to compare results and objectives. It may refer to discrepancies between sets of results, to experimental errors and to proposals for extending or improving the work undertaken. Comparisons may be made with other published work, which should be referenced.

#### *6.4.5.9 Conclusions (technical and non-technical)*

Conclusions should be specific, precise, pertinent and relate to what has (or has not) been achieved in the work. *The conclusions must not be based on surmise.*

#### *6.4.5.10 References (technical and non-technical)*

References referred to in the text should be listed here fully, e.g. for a book:

1 Todd, J. D. (1974) *Structural theory and analysis*, Macmillan, London, pp. 100–105.

and for a paper:

2 Smith, A. T. (1970) Phase calibration of displacement transducers. *Proceedings of Institution of Mechanical Engineers*, Vol. 107, No. 45, pp. 127–135 (January 1970).

#### *6.4.5.11 Appendices (technical and non-technical)*

These should be used for inessential information, for example computer printouts and hard copy, ancillary technical data, case-study material, etc.

## 6.5 Checklist

The following may help you in evaluating whether your report is likely to be satisfactory:

- Ensure that the report includes the following (where appropriate):

| | (tick) |
|---|---|
| Title | ☐ |
| Summary | ☐ |
| List of Contents | ☐ |
| Introduction | ☐ |
| Theory | ☐ |
| Apparatus | ☐ |
| Experimental method | ☐ |
| Background | ☐ |
| Methodology or Method of Investigation or Approach | ☐ |
| Analysis | ☐ |
| Results | ☐ |
| Discussion | ☐ |
| Conclusions | ☐ |
| References | ☐ |
| Appendices | ☐ |

- Does the summary clearly state the scope and results of the work?
- Does the introduction state the object of the work?
- Does the theory section contain only relevant theory? Is that theory used in the results or discussion section?
- Are the results clearly presented in graphs, tables or other form? Are all tables and figures numbered and referred to in the text?
- Are the results discussed critically? Are their implications clearly thought through?
- Are specific, clear conclusions drawn?
- Does the list of references contain only books or papers referred to specifically in the text?
- Is the writing style clear and concise? Are all sentences properly constructed?
- Has all unnecessary algebraic or arithmetic working in the text been removed?
- Is the main body of the report less than 2000 words?
- Does the report clearly state what you did, what happened and what the results mean?

# Part Two
# Quick Reference Guide

## An introduction to the Quick Reference Guide

The 'Quick Reference Guide' contains more than just the engineering information required for the projects in this book. It is intended to be a long-term source of useful engineering reference material which you will use for later design and manufacturing studies and as a practising engineer.

Once you have carried out most of the types of project illustrated in the main text, it is likely that you will have to develop more complex forms of engineering design for which you will require useful basic reference knowledge regarding material properties, mathematical formulae, design methods, manufacturing technology, etc. The comprehensive outline of general engineering information given in this section should meet most of your needs.

# 7
# Hardware

## 7.1 Bearings

### *7.1.1 Introduction*

Rolling element bearings are often referred to as ball bearings or ball races, but their correct title is an attempt to show that balls are not the only rolling shapes; other types are cylindrical rollers, tapered rollers and spherical rollers (see diagram).

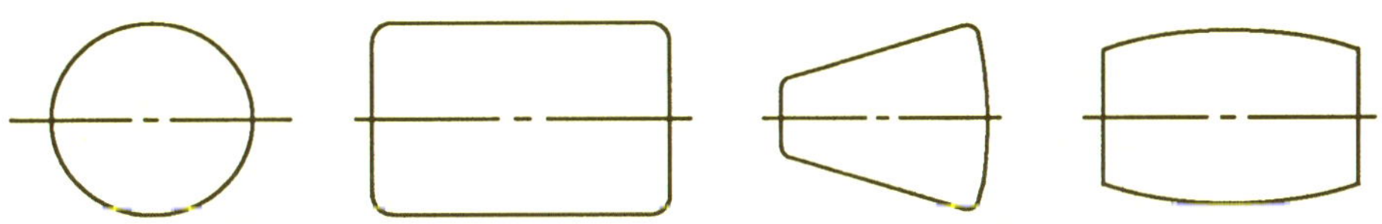


Double row radial ball bearings are available when increased load-carrying capacity is required but are not considered here. Similarly, although the bearings featured each carry axial thrust loads as well as radial loads, thrust bearings as such are not included; for details refer to manufacturers' catalogues.

### *7.1.2 Some useful rules of thumb*

1 The 'rules' and relationships given in this section are approximations which are useful when drafting a bearing assembly arrangement. They are sufficiently accurate for non-critical applications but for more demanding designs reference should be made to manufacturers' catalogues.
2 Single row radial ball bearings are available in a number of series, e.g. extra light, light, medium and heavy; while even smaller (miniature) bearings are available. Angular contact bearings may not be available in a 'heavy' series; roller bearings may only be available in 'light' and 'medium' series.
3 Single row radial bearings – whether ball, angular contact or roller – have the same overall sizes within any series.
4 (a) The outside dimensions of extra light ball bearings are generally 20–25% smaller than the medium series equivalents: their load carrying capability is about 45% less.
(b) Light ball bearing are generally 10–20% smaller than the medium series equivalents; their load-carrying capability is about 35% less.

(c) Heavy series ball bearings are generally 25–40% bigger than the medium series equivalents; their load-carrying capability is up to 90% higher.

5 (a) The radial load-carrying capability of angular contact bearings is about 15–20% higher than single row radial ball bearings.

(b) Cylindrical roller bearings can carry radial loads 50–90% greater than a ball bearing of the same overall dimensions.

6 (a) Single row radial ball bearings can carry axial loads, in either direction, up to 70% of their radial load-carrying capability.

(b) Although angular contact bearings are specially designed to carry substantial axial, as well as radial, loads they can do so in one direction only. When axial loads alternate in direction, bearings must be used in pairs, situated side-by-side, back-to-back or face-to-face.

(c) Standard roller bearings are not designed to carry axial loads but special designs for axial components of load are available from leading manufacturers.

### *7.1.3 Dimensions*

Diagrams of each type of bearing are now presented. Table 7.1 summarizes their performance characteristics for radial load, axial load, speed and tolerance of misalignment.

**Table 7.1** *Bearing performance characteristics*

| *Type* | *Radial load* | *Axial load* | *Speed* | *Tolerance of misalignment* |
|---|---|---|---|---|
| SRRBB | Very good | Good | Excellent | Good |
| SRACBB | Very good | Very good | Very good | Poor/Good |
| SRCRB | Excellent | – | Very good | Poor |

**Single Row Angular Contact Ball Bearings (SRACBB)**

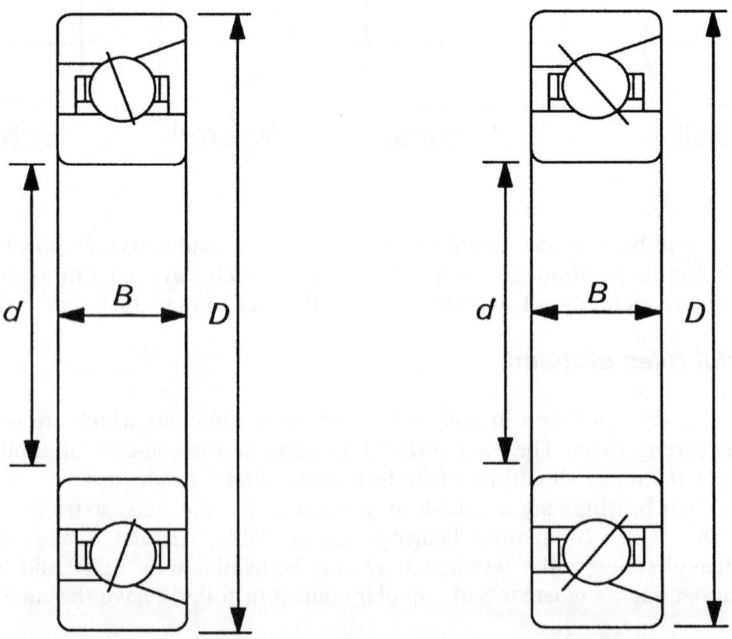


20° Contact Angle     40° Contact Angle

## Single Row Radial Ball Bearings (SRRBB)

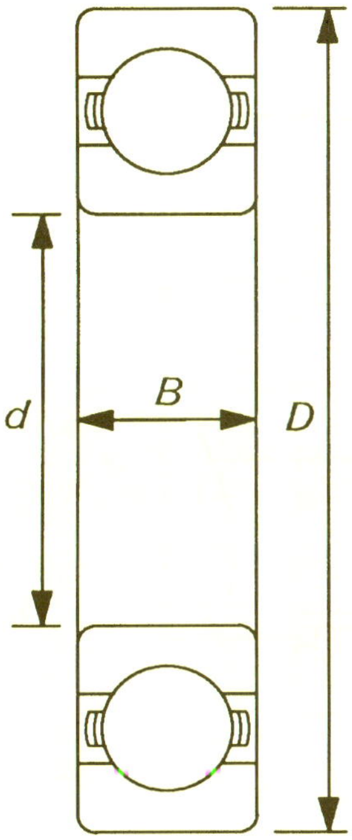


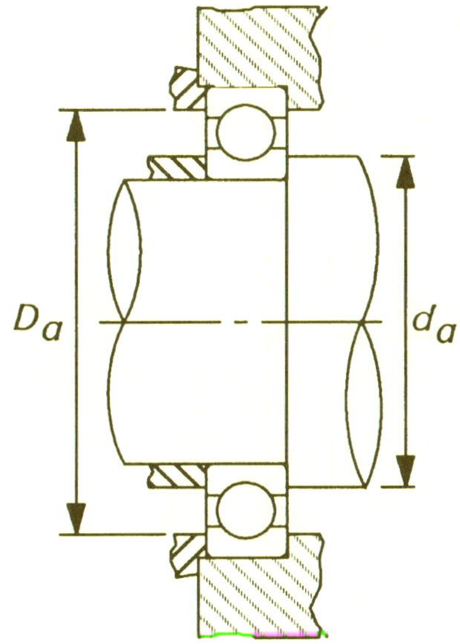


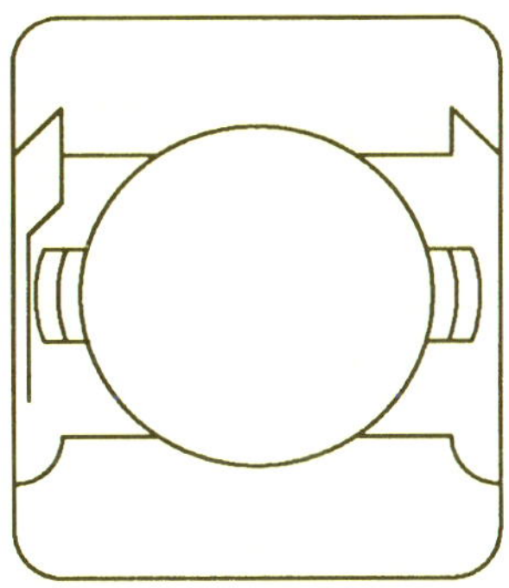

Single Seal

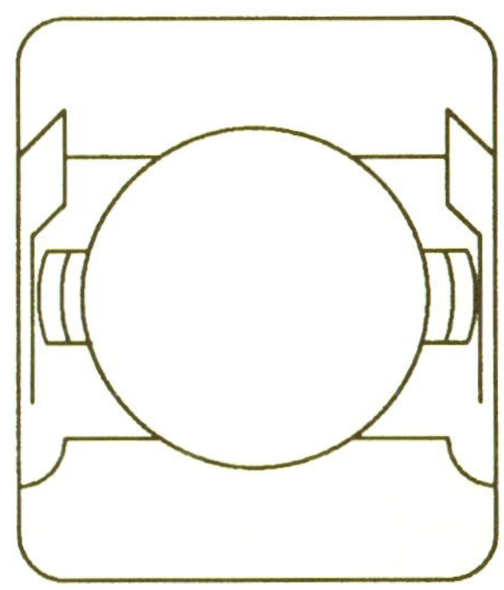

Double Seal

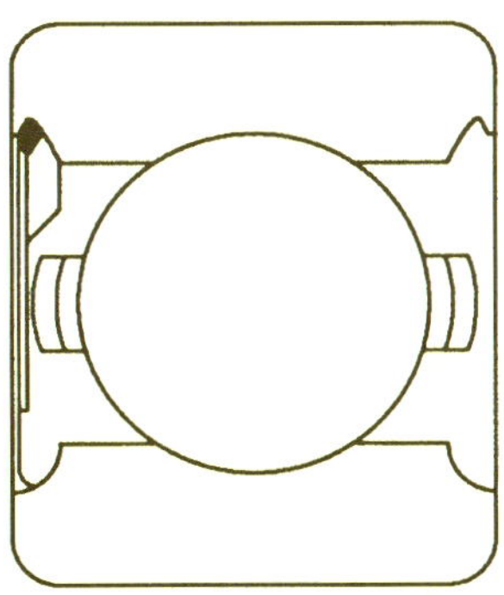

Single Shield

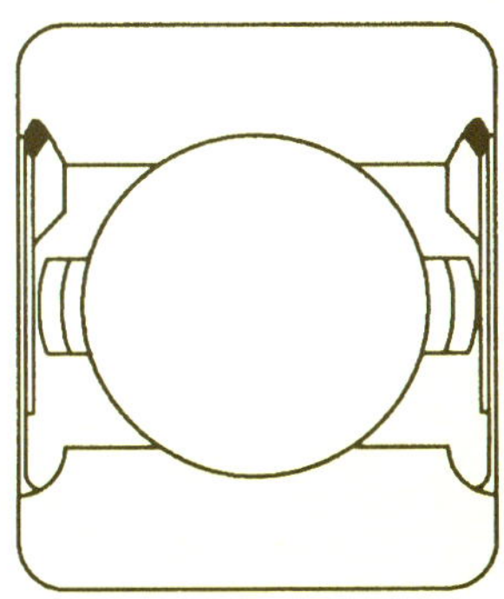

Double Shield

N.B. Seal + shield also possible

## Single Row Cylindrical Roller Bearings (SRCRB)

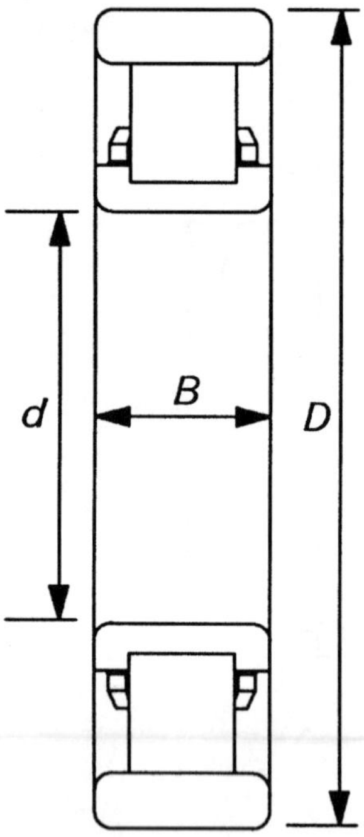


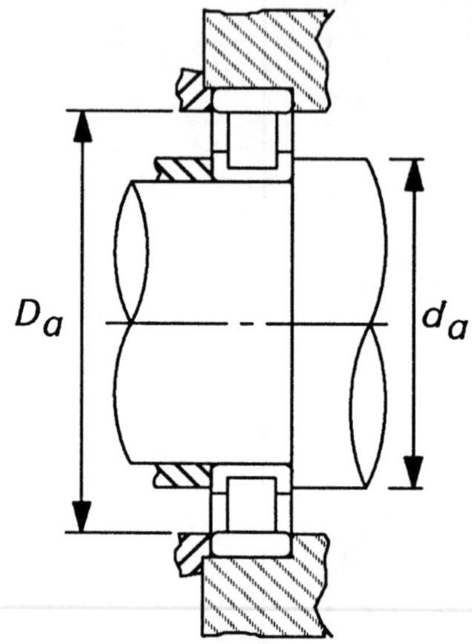


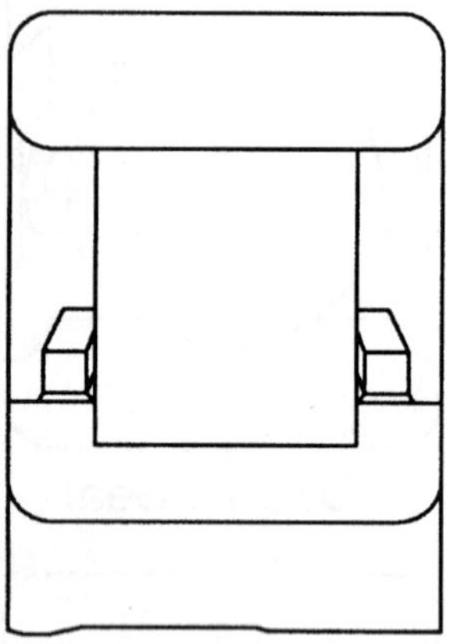

Loose Outer Ring

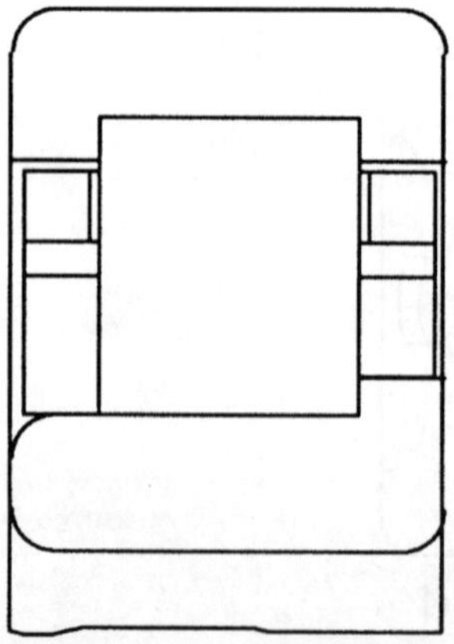

Flanged Inner Ring
(for unidirectional axial load)

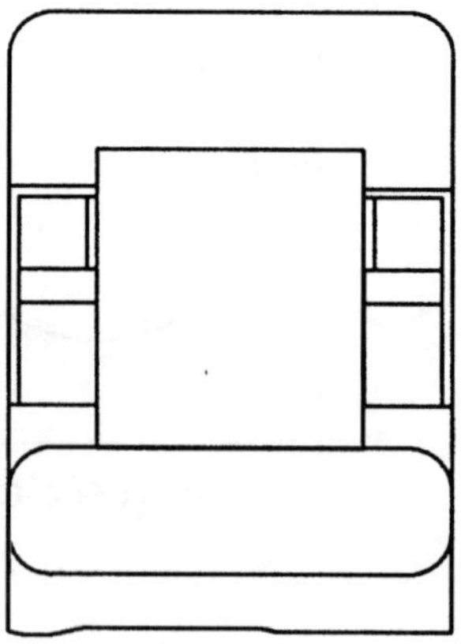

Loose Inner Ring

*Medium series dimensions (see figures for dimensions)*

| *d* (mm) | *D* (mm) | *B* (mm) | $d_a$ (mm) | | $D_a$ (mm) | | Fillet radius** (mm) |
|---|---|---|---|---|---|---|---|
| | | | *min** | *max** | *min** | *max** | |
| 10 | 35 | 11 | 12.4 | 15.5 | 30.0 | 32.6 | 0.6 |
| 12 | 37 | 12 | 16.0 | 16.5 | 32.5 | 33.0 | 1.0 |
| 15 | 42 | 13 | 19.0 | 20.1 | 37.6 | 38.0 | 1.0 |
| 17 | 47 | 14 | 21.0 | 22.9 | 41.9 | 43.0 | 1.0 |
| 20 | 52 | 15 | 24.0 | 26.2 | 45.2 | 48.0 | 1.0 |
| 25 | 62 | 17 | 29.0 | 33.0 | 55.1 | 58.0 | 1.0 |
| 30 | 72 | 19 | 34.0 | 39.1 | 64.5 | 68.0 | 1.0 |
| 35 | 80 | 21 | 41.0 | 43.9 | 71.1 | 74.0 | 1.5 |
| 40 | 90 | 23 | 46.0 | 49.5 | 81.3 | 84.0 | 1.5 |
| 45 | 100 | 25 | 51.0 | 55.1 | 89.9 | 94.0 | 1.5 |
| 50 | 110 | 27 | 58.0 | 61.0 | 99.1 | 102.0 | 2.0 |

*The sizes given are for ball bearings. Refer to the manufacturer's catalogue for sizes relating to angular contact and roller bearings.
**The fillet radius is the maximum allowable inside corner radius on the shaft shoulder and in the housing recess.

### 7.1.4 Fits for rolling element bearings

The bearing should be a sliding or interference fit on the shaft depending on whether the inner ring experiences a stationary or rotating load line.

Similarly, the bearing should be a sliding or interference fit in its housing depending on whether the outer ring is subjected to a stationary or rotating load line.

*Shaft seating limits for radial bearings to standard limits*

| Nominal shaft diameter (mm) | | Interference fit shaft limits (μm) | | | Sliding fit shaft limits (μm) | | |
|---|---|---|---|---|---|---|---|
| over | including | class | max. | min. | class | max. | min. |
| – | 6 | j5 | +3 | −2 | g6 | −4 | −12 |
| 6 | 10 | j5 | +4 | −2 | g6 | −5 | −14 |
| 10 | 18 | j5 | +5 | −3 | g6 | −6 | −17 |
| 18 | 30 | j5 | +5 | −4 | g6 | −7 | −20 |
| 30 | 50 | j5 | +6 | −5 | g6 | −9 | −25 |

*Housing seating limits for radial bearings to standard limits*

| Nominal housing bore (mm) | | Interference fit housing limits (μm) | | | Sliding fit housing limits (μm) | | |
|---|---|---|---|---|---|---|---|
| over | including | class | max. | min. | class | max. | min. |
| 30 | 50 | M6 | −4 | −20 | J6 | +10 | −6 |
| 50 | 80 | M6 | −5 | −24 | J6 | +13 | −6 |
| 80 | 120 | M6 | −6 | −28 | J6 | +16 | −6 |
| 120 | 150 | M6 | −8 | −33 | J6 | +18 | −7 |

### *7.1.5 Diagrams illustrating rotating and stationary load lines*

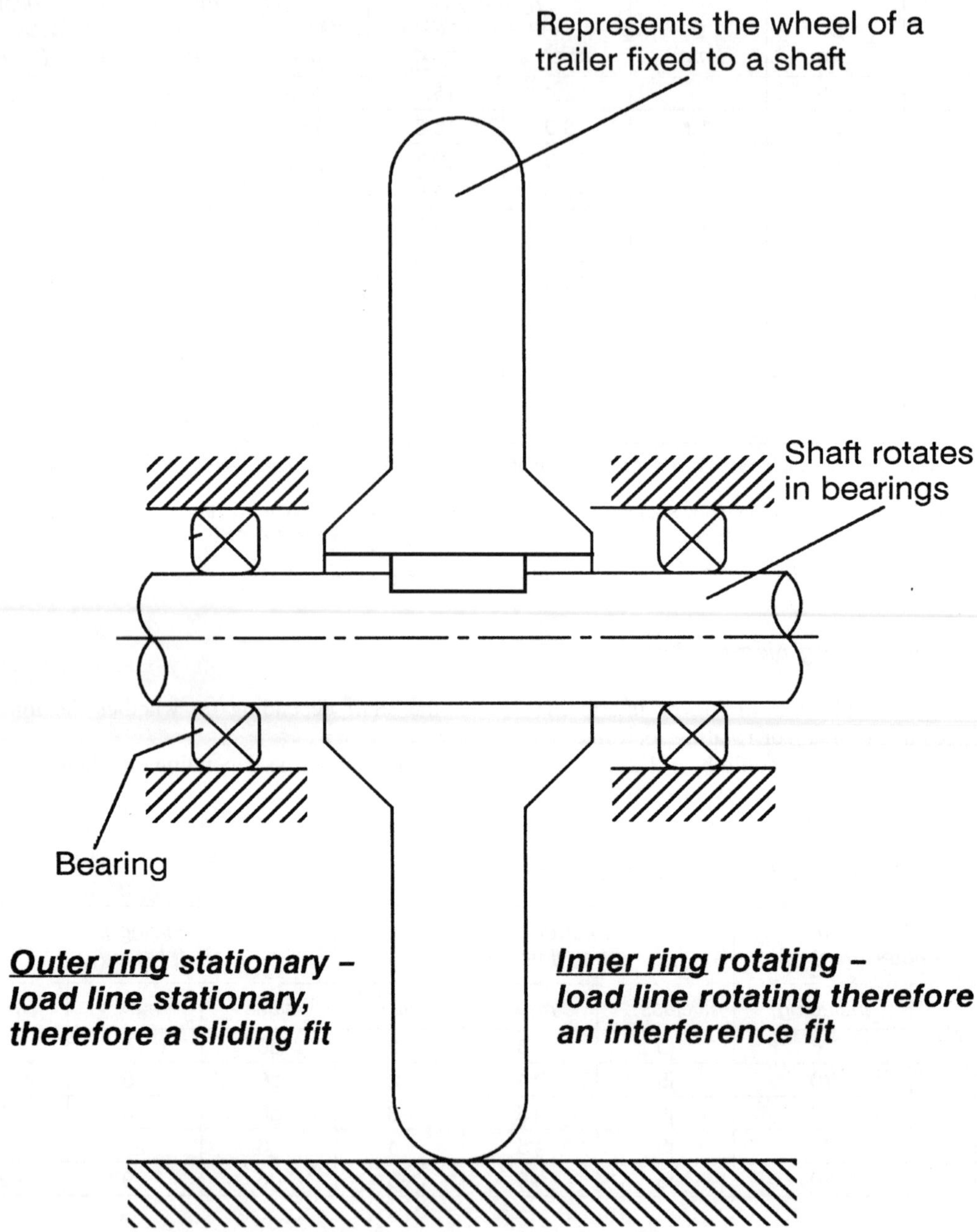

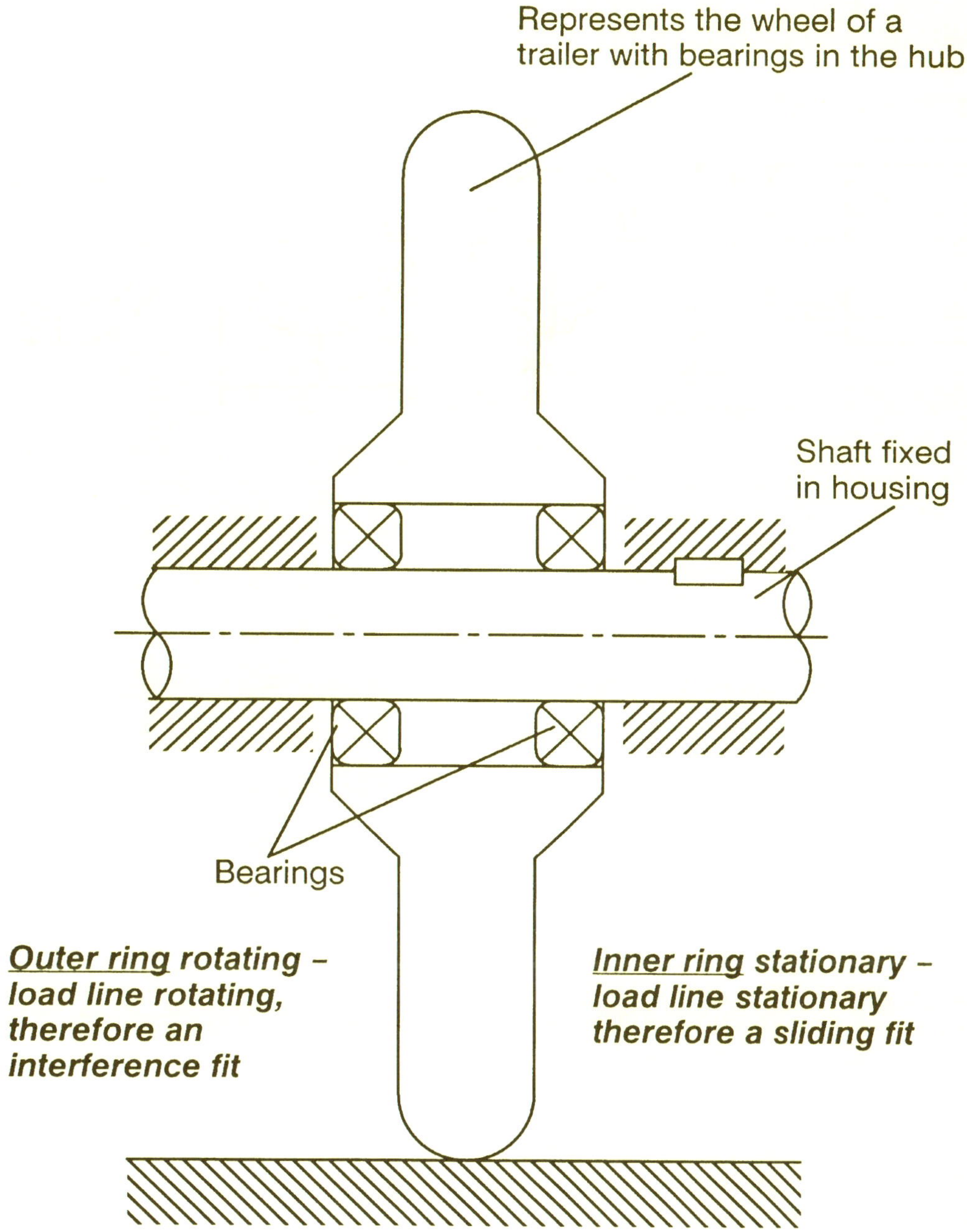
Represents the wheel of a trailer with bearings in the hub
Shaft fixed in housing
Bearings
Outer ring rotating – load line rotating, therefore an interference fit
Inner ring stationary – load line stationary therefore a sliding fit

## 7.2 Circlips

### *7.2.1 Standard external circlips – metric sizes*

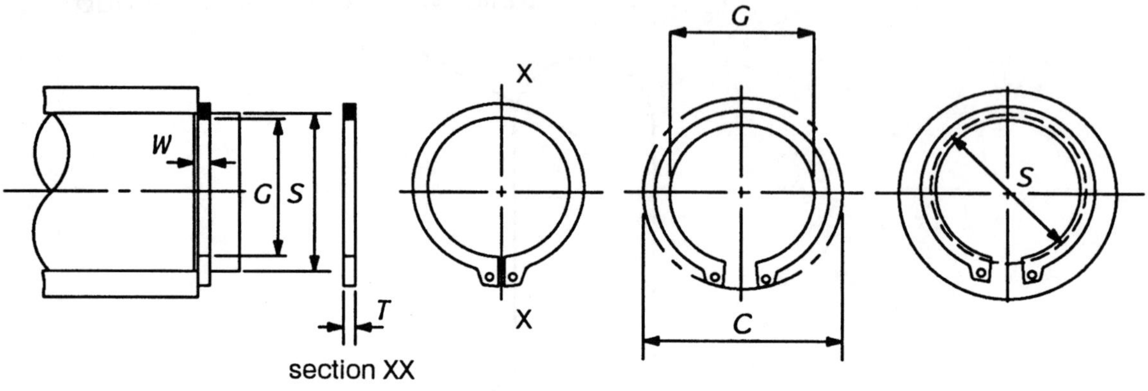

| Nominal shaft dia. S (mm) | Groove dimensions | | | Ring dimensions | | | | |
|---|---|---|---|---|---|---|---|---|
| | Diameter | | Width | Thickness | | Approx. clearance C (mm) | Thrust load (kg) Safety factor = 4 |
| | G (mm) | Tolerance (mm) | W min. (mm) | T (mm) | Tolerance (mm) | | |
| 3 | 2.8 | +0.000 −0.040 | 0.50 | 0.40 | +0.000 −0.060 | 7 | 68 |
| 4 | 3.8 | | | | | 8 | 90 |
| 5 | 4.8 | | 0.70 | 0.60 | | 10 | 176 |
| 6 | 5.7 | | 0.80 | 0.70 | | 12 | 252 |
| 7 | 6.7 | +0.000 −0.060 | 0.90 | 0.80 | | 14 | 341 |
| 8 | 7.6 | | | | | 15 | 390 |
| 9 | 8.6 | | 1.10 | 1.00 | | 16 | 560 |
| 10 | 9.6 | +0.000 −0.110 | | | | 17 | 622 |
| 11 | 10.5 | | | | | 18 | 685 |
| 12 | 11.5 | | | | | 19 | 747 |
| 13 | 12.4 | | | | | 20 | 809 |
| 14 | 13.4 | | | | | 22 | 872 |
| 15 | 14.3 | | | | | 23 | 935 |
| 16 | 15.2 | | | | | 24 | 996 |
| 17 | 16.2 | | | | | 25 | 1058 |
| 18 | 17.0 | | 1.30 | 1.20 | | 26 | 1684 |
| 19 | 18.0 | | | | | 27 | 1778 |
| 20 | 19.0 | +0.000 −0.150 | | | | 28 | 1872 |
| 21 | 20.0 | | | | | 30 | 1966 |
| 22 | 21.0 | | | | | 31 | 2059 |
| 23 | 22.0 | | | | | 32 | 2166 |
| 24 | 22.9 | +0.000 −0.210 | | | | 33 | 2247 |
| 25 | 23.9 | | | | | 34 | 2340 |
| 26 | 24.9 | | | | | 35 | 2433 |
| 27 | 25.6 | | | | | 36 | 2544 |
| 28 | 26.6 | | 1.60 | 1.50 | | 38 | 3298 |
| 29 | 27.6 | | | | | 39 | 3416 |
| 30 | 28.6 | | | | | 40 | 3534 |
| 31 | 29.3 | | | | | 41 | 3650 |
| 32 | 30.3 | +0.000 −0.250 | | | | 43 | 3769 |
| 33 | 31.3 | | | | | 44 | 3887 |
| 34 | 32.3 | | | | | 45 | 4005 |
| 35 | 33.0 | | | | | 46 | 4123 |
| 36 | 34.0 | | | | | 47 | 4998 |
| 37 | 35.0 | | | | | 49 | 5135 |
| 38 | 36.0 | | 1.85 | 1.75 | | 50 | 5276 |
| 39 | 37.0 | | | | | 51 | 5413 |
| 40 | 37.5 | | | | | 53 | 5553 |
| 41 | 38.5 | | | | | 54 | 5692 |
| 42 | 39.5 | | | | | 55 | 5831 |
| 44 | 41.5 | | | | | 57 | 6108 |
| 45 | 42.5 | | | | | 58 | 6247 |
| 46 | 43.5 | | | | | 59 | 6386 |
| 47 | 44.5 | | | | | 60 | 6524 |
| 48 | 45.5 | | | | | 62 | 6664 |
| 50 | 47.0 | | | | | 64 | 7993 |

### *7.2.2 Standard internal circlips – metric sizes*

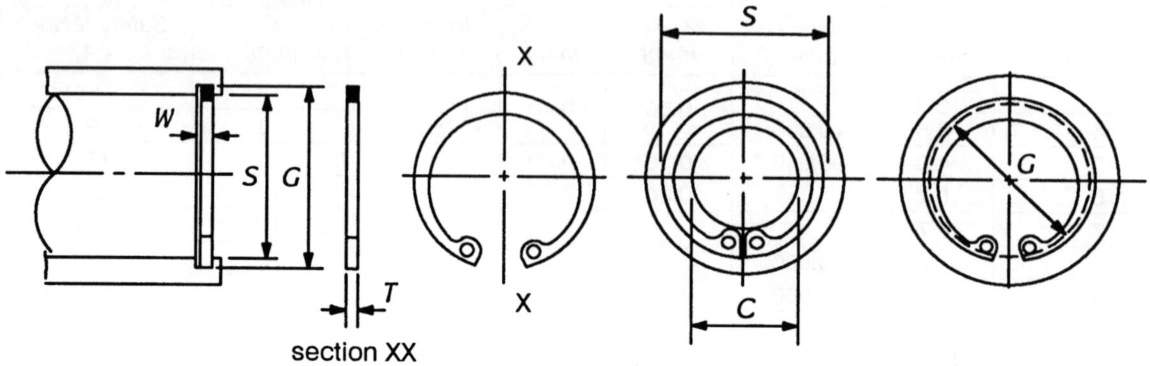

| Nominal housing dia. S (mm) | Groove dimensions: Diameter G (mm) | Groove dimensions: Diameter Tolerance (mm) | Groove dimensions: Width W min. (mm) | Ring dimensions: Thickness T (mm) | Ring dimensions: Thickness Tolerance (mm) | Approx. clearance C (mm) | Thrust load (kg) Safety factor = 4 |
|---|---|---|---|---|---|---|---|
| 8 | 8.4 | +0.09 −0.00 | 0.90 | 0.80 | +0.00 −0.06 | 3.0 | 469 |
| 9 | 9.4 | | | | | 4.0 | 527 |
| 9.5 | 9.9 | +0.11 −0.00 | 1.10 | 1.00 | | 3.0 | 710 |
| 10 | 10.4 | | | | | 3.0 | 747 |
| 10.5 | 10.9 | | | | | 3.0 | 784 |
| 11 | 11.4 | | | | | 4.0 | 822 |
| 12 | 12.5 | | | | | 5.0 | 897 |
| 13 | 13.6 | | | | | 6.0 | 971 |
| 14 | 14.6 | | | | | 7.0 | 1046 |
| 15 | 15.7 | | | | | 8.0 | 1121 |
| 16 | 16.8 | | | | | 8.0 | 1195 |
| 17 | 17.8 | | | | | 9.0 | 1270 |
| 18 | 19.0 | +0.15 −0.00 | | | | 10.0 | 1345 |
| 19 | 20.0 | | | | | 11.0 | 1420 |
| 20 | 21.0 | | | | | 12.0 | 1494 |
| 21 | 22.0 | | | | | 12.0 | 1569 |
| 22 | 23.0 | | | | | 13.0 | 1644 |
| 23 | 24.1 | | 1.30 | 1.20 | | 15.0 | 2600 |
| 24 | 25.2 | +0.21 −0.00 | | | | 15.0 | 2696 |
| 25 | 26.2 | | | | | 16.0 | 2808 |
| 26 | 27.2 | | | | | 16.0 | 2920 |
| 27 | 28.4 | | | | | 18.0 | 3053 |
| 28 | 29.4 | | | | | 18.0 | 3145 |
| 29 | 30.4 | +0.25 −0.00 | | | | 20.0 | 3258 |
| 30 | 31.4 | | | | | 20.0 | 3370 |
| 31 | 32.7 | | | | | 21.0 | 3370 |
| 32 | 33.7 | | | | | 21.0 | 3595 |
| 33 | 34.7 | | | | | 22.0 | 2595 |
| 34 | 35.7 | | 1.60 | 1.50 | | 23.0 | 4806 |
| 35 | 37.0 | | | | | 24.0 | 4948 |
| 36 | 38.0 | | | | | 25.0 | 5089 |
| 37 | 39.0 | | | | | 26.0 | 5089 |
| 38 | 40.0 | | | | | 27.0 | 5372 |
| 39 | 41.0 | | | | | 28.0 | 5372 |
| 40 | 42.5 | | 1.85 | 1.75 | | 28.0 | 6664 |
| 41 | 43.5 | | | | | 29.0 | 6831 |
| 42 | 44.5 | | | | | 30.0 | 6998 |
| 43 | 45.5 | | | | | 31.0 | 7165 |
| 44 | 46.5 | | | | | 31.0 | 7330 |
| 45 | 47.5 | | | | | 33.0 | 7497 |
| 46 | 48.5 | | | | | 34.0 | 7664 |
| 47 | 49.5 | | | | | 34.0 | 7831 |
| 48 | 50.5 | | | | | 35.0 | 7997 |

## 7.3 'Pop'™ rivets

'Pop'™ rivets are to blind rivets what 'Hoover' is to vacuum cleaners. These types of rivets allow two or more materials to be permanently joined together from one side: access to the other 'blind' side is not necessary. The materials are held in position and a hole drilled through them; the rivet is then inserted and expanded on the blind side by pulling the headed pin (or mandrel) with a special rivet tool until the pin snaps off at a pre-determined load.

'Pop'™ rivets are available in a variety of diameters, lengths and materials. There are also variants designed for special applications: the sealed type are water and pressure tight; the peel type allow materials

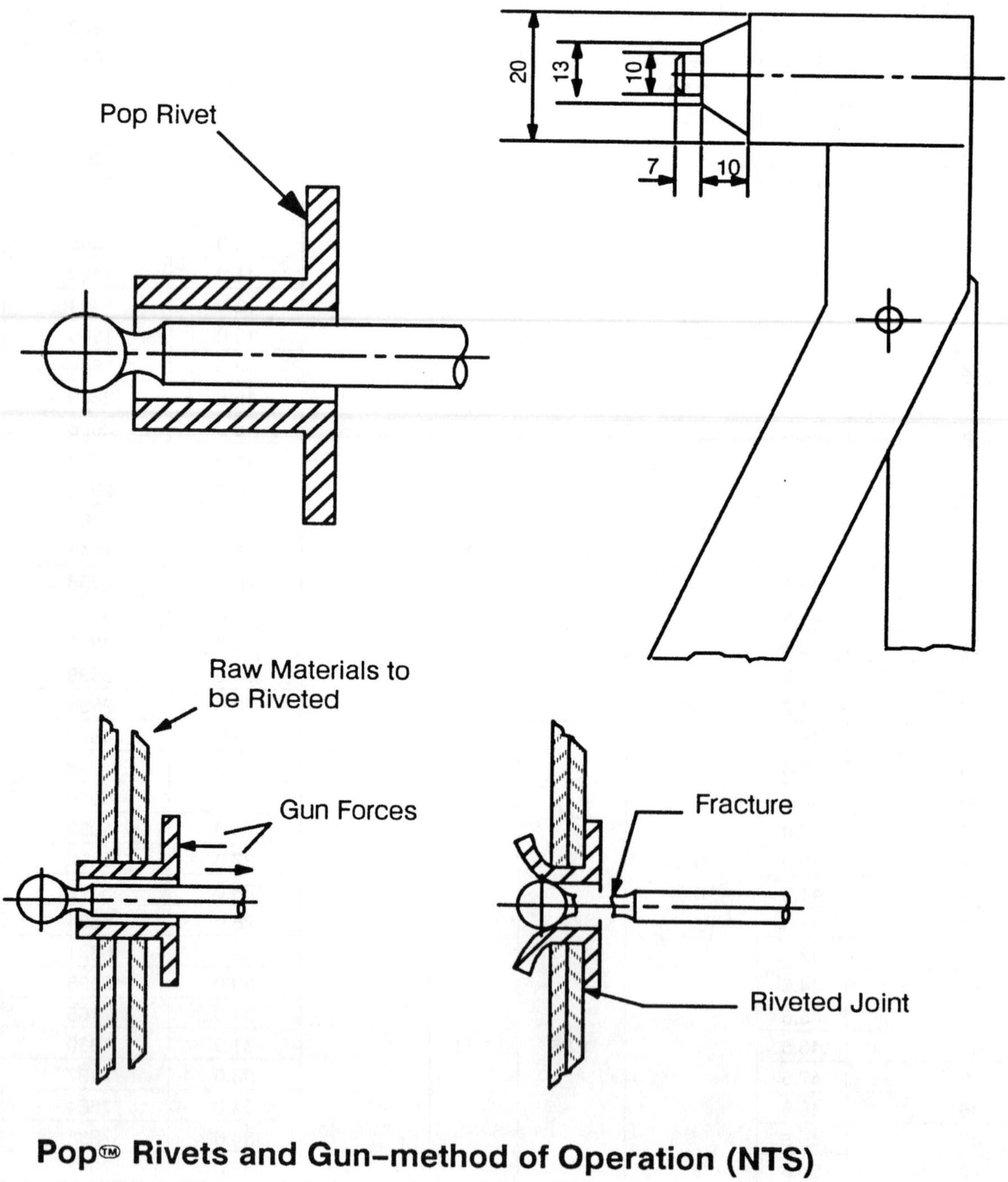


**Pop™ Rivets and Gun–method of Operation (NTS)**

| *Nominal rivet diameter mm (in.)* | *Maximum riveting thickness* | | | | *Nominal rivet* | | *Drill size (mm)* |
|---|---|---|---|---|---|---|---|
| | *Domed head* | | *120° Csk. head* | | *Body length* | | |
| | *mm* | *in.* | *mm* | *in.* | *mm* | *in.* | |
| | 1.6 | 0.062 | 2.7 | 0.105 | 4.3 | 0.170 | |
| | 3.2 | 0.125 | 4.3 | 0.168 | 6.1 | 0.241 | |
| | 4.8 | 0.187 | 5.8 | 0.230 | 7.9 | 0.310 | |
| | 6.4 | 0.250 | 7.4 | 0.293 | 9.7 | 0.382 | |
| | 7.9 | 0.312 | 9.0 | 0.355 | 11.5 | 0.451 | |
| 3.2 (0.125) | 9.5 | 0.375 | 10.6 | 0.418 | 13.3 | 0.522 | 3.30 |
| | 11.1 | 0.437 | 12.2 | 0.480 | 15.0 | 0.592 | |
| | 12.7 | 0.500 | 13.8 | 0.543 | 16.8 | 0.663 | |
| | 14.3 | 0.562 | 15.4 | 0.605 | 18.6 | 0.732 | |
| | 16.6 | 0.656 | 17.7 | 0.699 | 20.2 | 0.795 | |
| | 19.8 | 0.781 | 20.9 | 0.824 | 24.2 | 0.951 | |
| | 3.2 | 0.125 | 4.5 | 0.178 | 6.8 | 0.266 | |
| | 4.8 | 0.187 | 6.1 | 0.240 | 8.5 | 0.335 | |
| | 6.4 | 0.250 | 7.7 | 0.303 | 10.3 | 0.406 | |
| | 7.9 | 0.312 | 9.3 | 0.365 | 12.1 | 0.476 | |
| | 9.5 | 0.375 | 10.9 | 0.428 | 13.9 | 0.547 | |
| 4.0 (0.156) | 11.1 | 0.437 | 12.4 | 0.490 | 15.7 | 0.617 | 0.617 |
| | 12.7 | 0.500 | 14.0 | 0.553 | 17.4 | 0.689 | |
| | 13.5 | 0.532 | 14.8 | 0.585 | 18.3 | 0.722 | |
| | 15.9 | 0.625 | 17.2 | 0.678 | 20.2 | 0.795 | |
| | 17.4 | 0.687 | 18.8 | 0.740 | 22.3 | 0.877 | |
| | 19.8 | 0.781 | 21.2 | 0.834 | 24.7 | 0.971 | |
| | 3.2 | 0.125 | 4.8 | 0.188 | 7.4 | 0.290 | |
| | 4.8 | 0.187 | 6.4 | 0.250 | 9.2 | 0.361 | |
| | 6.4 | 0.250 | 8.0 | 0.313 | 10.9 | 0.431 | |
| | 7.9 | 0.312 | 9.5 | 0.375 | 12.7 | 0.501 | |
| 4.8 (0.187) | 9.5 | 0.375 | 11.1 | 0.438 | 14.5 | 0.572 | 4.90 |
| | 11.1 | 0.437 | 12.7 | 0.500 | 16.3 | 0.641 | |
| | 13.5 | 0.532 | 15.1 | 0.595 | 19.0 | 0.748 | |
| | 19.8 | 0.781 | 21.4 | 0.844 | 25.4 | 1.000 | |
| | 26.2 | 1.031 | 27.3 | 1.094 | 31.8 | 1.250 | |
| | 6.4 | 0.250 | – | – | 12.7 | 0.500 | |
| 6.4 (0.250) | 12.7 | 0.500 | – | – | 19.5 | 0.766 | 6.50 |
| | 19.0 | 0.750 | – | – | 26.2 | 1.032 | |

such as plastics, rubbers and plywoods to be joined to metal frames and panels; the grooved type are designed to expand inside rather than behind materials such as hardboard, plywood, glass fibre and concrete.

'Pop'™ is a trade mark of Tucker Fasteners Ltd.

Note that the drill size is usually the nominal diameter plus 0.1 mm.

Common materials for rivets of this kind are aluminium/magnesium alloys, mild steel, stainless steel, copper and nickel/copper alloys. Note that all the rivet sizes tabulated here are available in all the materials listed.

## 7.4 Roll pins – metric sizes

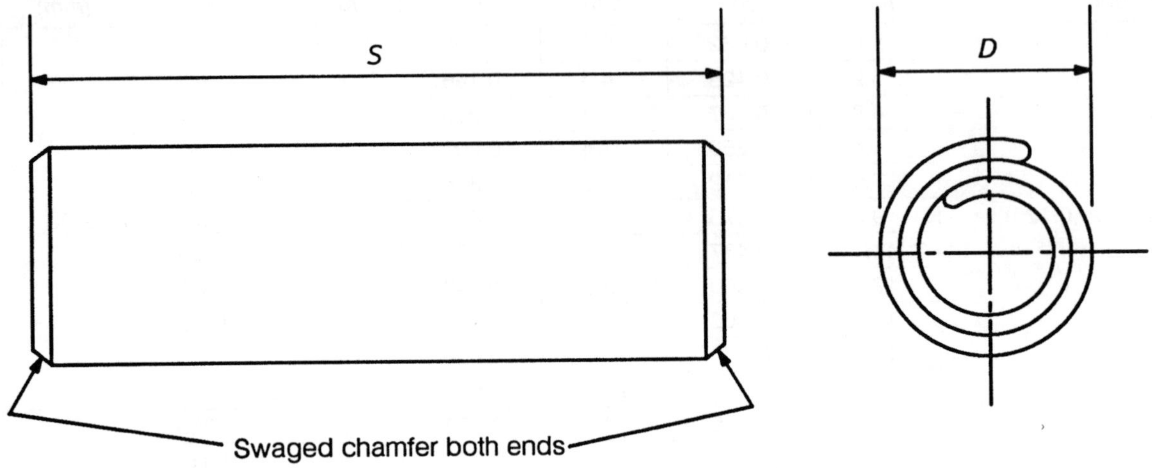


Roll pins or, as they are also commonly called, coiled pins are designed to be installed in holes slightly smaller in diameter than the pin itself. The coiled construction allows the pin to be compressed to suit the hole. Once in position the pin grips the side of the hole and cannot be shaken out by vibrations or other impacts. It can, however, be driven out by a pin punch and reused indefinitely.

Their principal advantage is that they can be used in ordinary drilled holes; no reaming or other hole finishing is necessary.

Roll pins are available in a number of different sizes and materials and for heavy duty, standard duty and light duty applications. The sizes given here are a selection of the most commonly used; the strength values quoted are for carbon alloy steel, standard duty pins.

| *Nominal diameter** *D (mm)* | *Overall length* *L (mm)* | *Double shear strength* *(kN)* |
|---|---|---|
| 1 | 12 | 0.6 |
| 1.5 | 12, 16 | 1.45 |
| 2 | 12, 20 | 2.5 |
| 3 | 12, 16, 24, 30 | 5.5 |
| 4 | 12, 16, 24, 40 | 9.6 |
| 5 | 12, 20, 30, 45 | 15.0 |
| 6 | 16, 25, 45, 55 | 22.0 |

* Holes may be drilled to the nominal size, i.e. for a 3 mm pin use a 3 mm

## 7.5 Springs

### 7.5.1 Spring selection line charts

The line charts presented in this section allow coil springs to be selected for particular applications. They can also be used to design springs but when only a few springs are required it is much more cost effective to choose a suitable spring from the manufacturer's standard range of products. The charts apply to both tension and compression cylindrical helical springs made from round steel wire.

Selection parameters:

(a) $S$ (N/mm) – spring stiffness
(b) $d$ (mm) – wire diameter *Note:* Many springs are still made from wires produced to standard wire gauges – refer to Sheet metal gauge comparisons, Section 8.6 of this Quick Reference Guide)
(c) $n$ – number of active turns (*Note:* Compression springs usually have closed ends ground square. Thus one or two turns do not contribute to the 'springiness' of the spring)
(d) $D$ (mm) – the mean spring diameter
(e) $C = D/d$ – a ratio usually between about 6 and 15
(f) $W$ (N) – the load on the spring.

### 7.5.2 Selection procedure

Usually the stiffness of the spring and the space available to accommodate it are the starting points. Chart 1 can be used on a 'try and see' basis. For example, a spring of stiffness between 2 and 3 N/mm may be required to fit in a space which will allow a mean diameter of 12 mm. Try a value for $C$ of around 10; this implies a value of 1.2 mm for $d$. Use a straight edge to join $C$ = 10 mm on the second scale to $d$ = 1.2 mm on the third scale; make a pencil mark where it crosses the support line. Place the straight edge with 2 N/mm on the left-hand scale and the pencil mark on the support line and read off the number of turns (6 in this case) from the right-hand scale. Repeat with 3 N/mm on the left-hand scale reading 4 from the right-hand scale. Note that the closed length of the first spring will be about $(6 + 2)d$ = 9.6 mm and the second $(4 + 2)d$ = 7.2 mm.

Suppose from the springs available the two nearest to the sizes assumed above are selected and found to be:

$d$ = 1.2 mm, $D$ = 10 mm, $n$ = 6

$d$ = 1.4 mm, $D$ = 12 mm, $n$ = 8

For the first spring, $C = D/d$ = 8.33 and the chart shows that for 6 active turns the stiffness is 3.5 N/mm. The compressed length will be about 9.6 mm.

For the second spring, $C = D/d$ = 8.57 and the chart shows that for 10 active turns the stiffness is about 2.8 N/mm. The compressed length will be about $(8 + 2) \times 1.4$ = 14 mm.

### 7.5.3 Acceptable stress levels

Although the yield stress of spring steel wire may be as high as 1250 N/mm$^2$, springs stressed to near this level are liable to fail eventually by fatigue, especially if they are subjected to continually varying stress conditions. Applications involving stress levels in excess of 700 N/mm$^2$ should therefore be carefully scrutinized.

Stress for any given load can be checked using the second line chart. The support line is marked at its intersection with a line drawn from the value of $C$ to the magnitude of the load. A line extended from the appropriate wire diameter through the mark will indicate the maximum fibre stress (N/mm$^2$) on the left-hand scale.

For the springs considered above:

(i) $d$ = 1.2 mm, $D$ = 10 mm, $n$ = 6, $C$ = 8.33 if the maximum load is 20 N, the maximum stress is approximately 320 N/mm$^2$
(ii) $d$ = 1.4 mm, $D$ = 12 mm, $n$ = 8, $C$ = 8.57 if the maximum load is 20 N, the maximum stress is approximately 240 N/mm$^2$.

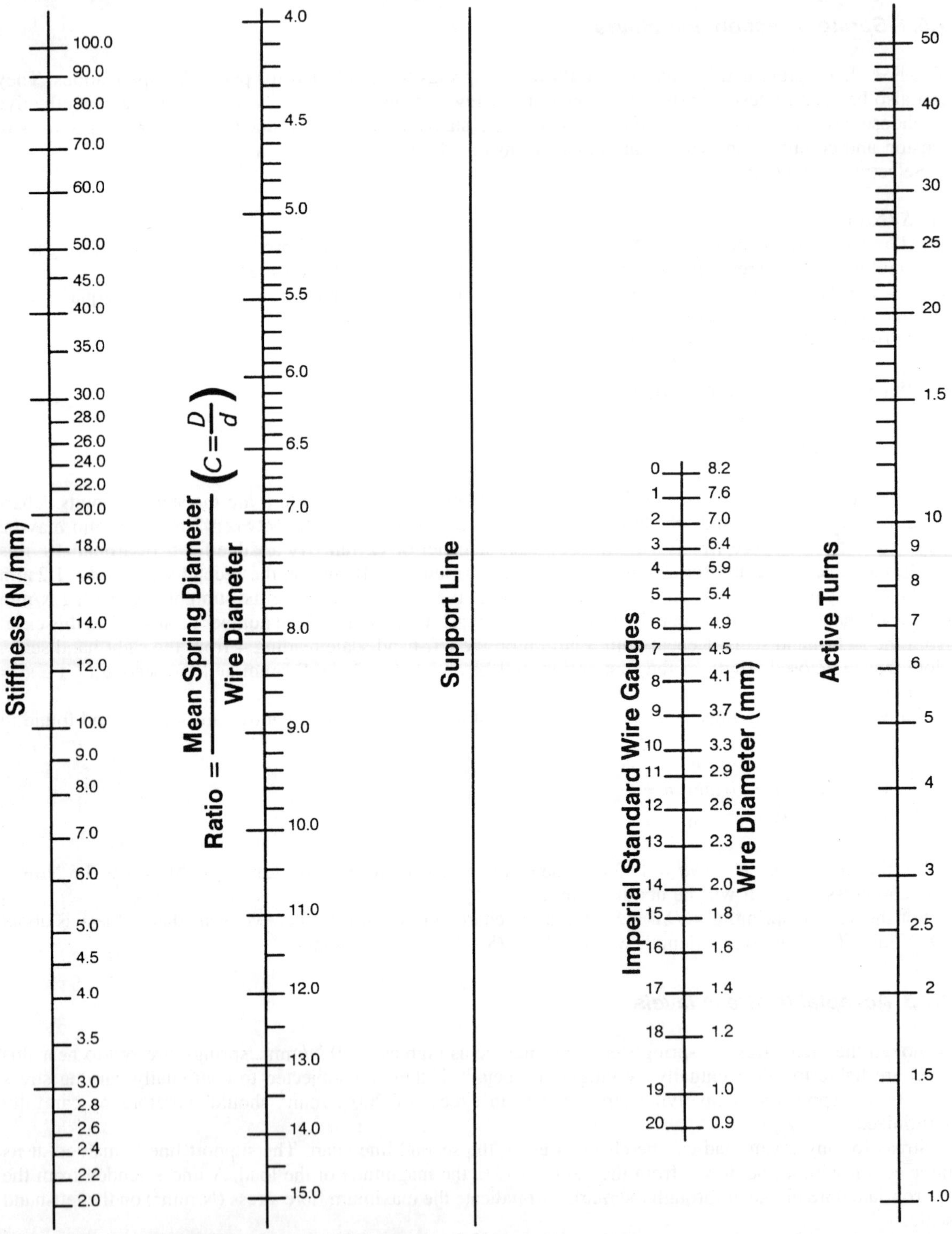


**Chart 1 – SPRING STIFFNESS**

**Line Chart for Round Wire Helical Tension and Compression Springs**

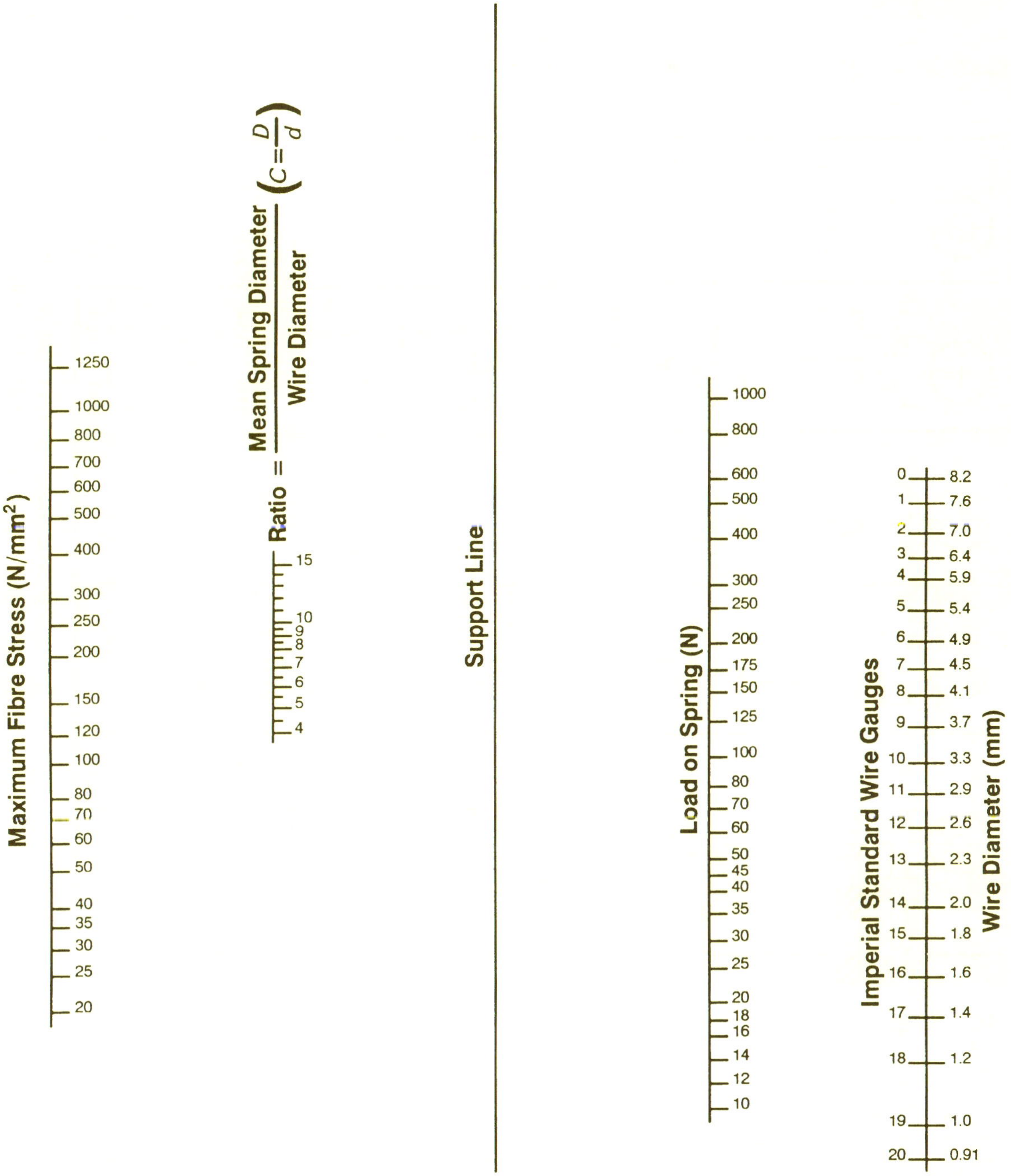


**Chart 2 – MAXIMUM FIBRE STRESS**

**Line Chart for Round Wire Helical Tension and Compression Springs**

## 7.6 Machine screws and nuts

### THREAD AND FASTENER KEY DIMENSIONS

*Note: The sizes given are for hexagonal socket styles. Dimensions for screwdriver styles, i.e. cheese head, countersink head and round head screws, may be slightly different but the sizes given in the tabulation are accurate enough for most practical purposes.*

**TYPE OF FASTENER** →

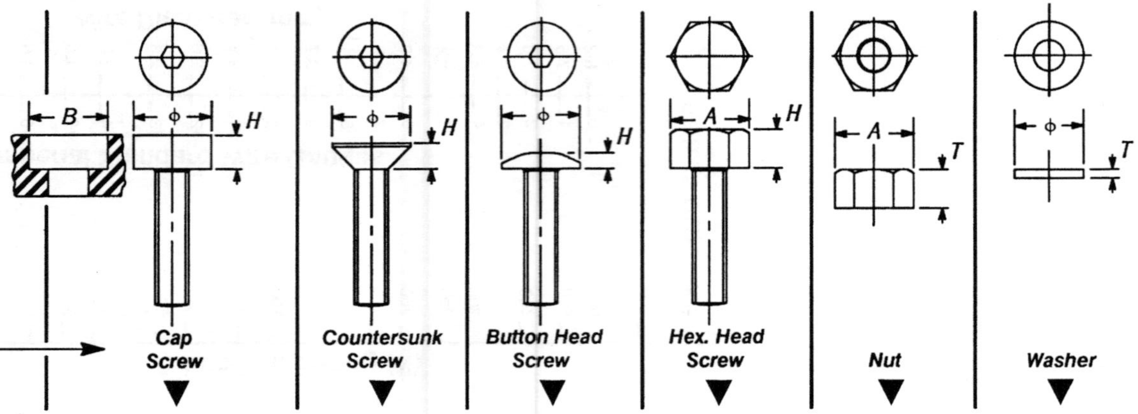


| Thread Size | Pitch (mm) | Tapping Drill Size (mm) | Clearance Drill Size (mm) | Cap Screw | | | Countersunk Screw | | Button Head Screw | | Hex. Head Screw | | Nut | | Washer | |
|---|---|---|---|---|---|---|---|---|---|---|---|---|---|---|---|---|
| | | | | B | ϕ | H | ϕ* | H* | ϕ | H | A | H | A | T | ϕ | T |
| M3 | 0.5 | 2.5 | 3.4 | 6.0 | 5.5 | 3.0 | 6.8 | 1.9 | 5.5 | 1.6 | – | – | – | – | 7.0 | 0.5 |
| M4 | 0.7 | 3.3 | 4.5 | 8.0 | 7.0 | 4.0 | 9.0 | 2.5 | 7.5 | 2.1 | – | – | – | – | 9.0 | 0.8 |
| M5 | 0.8 | 4.2 | 5.5 | 10.0 | 8.5 | 5.0 | 11.2 | 3.1 | 9.5 | 2.7 | 8.0 | 3.8 | 8.0 | 4.3 | 10.0 | 1.0 |
| M6 | 1.0 | 5.0 | 6.6 | 11.0 | 10.0 | 6.0 | 13.5 | 3.8 | 10.5 | 3.2 | 10.0 | 4.3 | 10.0 | 5.3 | 12.5 | 1.6 |
| M8 | 1.25 | 6.8 | 9.0 | 15.0 | 13.0 | 8.0 | 18.0 | 5.0 | 14.0 | 4.3 | 13.0 | 5.8 | 13.0 | 6.8 | 17.0 | 1.6 |
| M10 | 1.5 | 8.5 | 11.0 | 18.0 | 16.0 | 10.0 | 22.4 | 6.2 | 18.0 | 5.3 | 17.0 | 7.4 | 17.0 | 8.4 | 21.0 | 2.0 |
| M12 | 1.75 | 10.2 | 14.0 | 20.0 | 18.0 | 12.0 | 26.9 | 7.5 | 21.0 | 6.4 | 19.0 | 8.4 | 19.0 | 10.4 | 24.0 | 2.3 |
| (M14) | 2.0 | 12.0 | 16.0 | 24.0 | 21.0 | 14.0 | 30.3 | 8.2 | – | – | – | – | – | – | 28.0 | 2.5 |
| M16 | 2.0 | 14.0 | 18.0 | 26.0 | 24.0 | 16.0 | 33.6 | 8.8 | 28.0 | 8.8 | 24.0 | 10.4 | 24.0 | 13.5 | 30.0 | 3.0 |
| (M18) | 2.5 | 15.5 | 20.0 | 30.0 | 27.0 | 18.0 | 37.0 | 9.5 | – | – | – | – | – | – | 34.0 | 3.0 |
| M20 | 2.5 | 17.5 | 22.0 | 33.0 | 30.0 | 20.0 | 40.3 | 10.2 | – | – | 30.0 | 13.9 | 30.0 | 16.5 | 37.0 | 3.0 |
| (M22) | 2.5 | 19.5 | 24.0 | 36.0 | 33.0 | 22.0 | – | – | – | – | 32.0 | 14.9 | 32.0 | 18.5 | 39.0 | 3.0 |
| M24 | 3.0 | 21.0 | 26.0 | 40.0 | 36.0 | 24.0 | – | – | – | – | 36.0 | 15.9 | 36.0 | 19.6 | 44.0 | 4.0 |

Sizes shown in brackets, e.g. (M14) are non-preferred.

* The sizes given above are for the countersink hole so that screw heads do not protrude beyond the surface.

# 8 Manufacturing technology

## 8.1 Cutting tools, machining speeds and feeds and machining times

### *8.1.1 Introduction*

A machine tool can be defined as a power-driven machine used for producing a variety of shapes and sizes by cutting away excess material.

It has three main functions:

(a) to hold and control the material being cut
(b) to support and control the cutting tools used to remove the excess metal.
(c) to provide (cutting) motion between the material and the tool.

Three main types of cutting tool are used on machine tools:

(a) single-point tools
(b) multi-point tools
(c) abrasive wheels e.g. grinding machines (these are not covered in this guide).

### *8.1.2 Single-point tools*

A typical single-point tool is shown here, with a description of the key parameters and features. Single-point tools are cutting tools which have one cutting part and one shank.

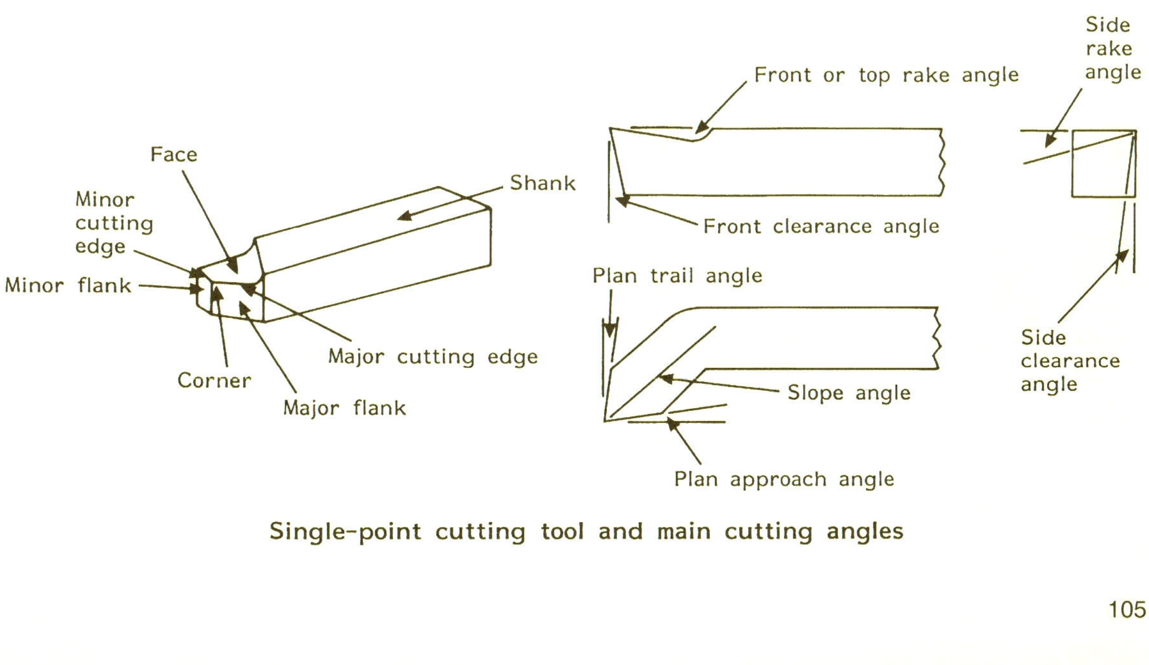


Single-point cutting tool and main cutting angles

The main features of such tools are:

(a) the *face* – the surface over which the chip or cut material flows
(b) the *flank* – the tool surfaces over which the workpiece passes
(c) the *cutting edge* – the edge which cuts the material; major cutting is carried out by the major cutting edge.

The sizes of the tool angles vary for different materials as shown in the table; note that these are for turning.

*High-speed steel tool geometry for typical materials*

| *Material* | *Front rake (degrees)* | *Front clearance (degrees)* | *Side rake (degrees)* | *Side clearance (degrees)* |
|---|---|---|---|---|
| Mild steel | 8–10 | 8 | 12–15 | 6 |
| Stainless steel | 5 | 8 | 6–10 | 6 |
| Cast iron | 8 | 6 | 10–15 | 6 |
| Brass/bronze | 1 | 8 | 0–3 | 6 |
| Aluminium | 35–55 | 6 | 10–20 | 1–3 |

Some basic single-point tool cutting actions are shown in the figure.

**Single-point tool cutting applications**

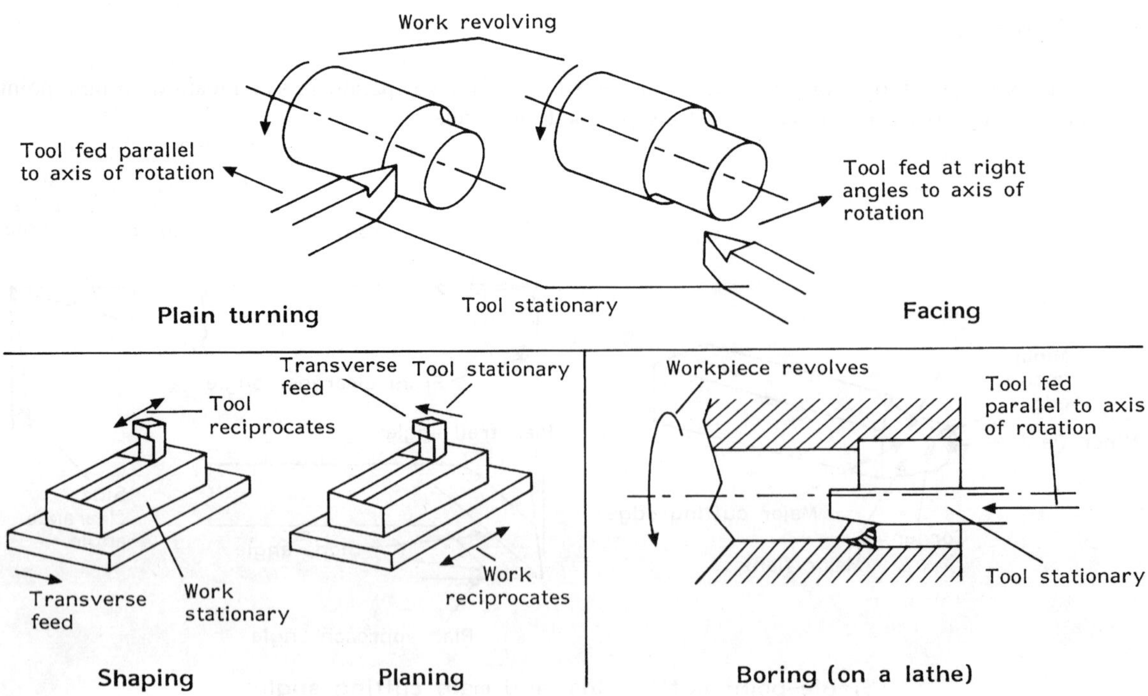

### 8.1.3 Multi-point tools

A multi-point tool is a series of two or more cutting edges secured to a common body. The vast majority of these types of cutters are intended to be rotated and have a shank which is inserted into a machine drive shaft.

A multi-point tool

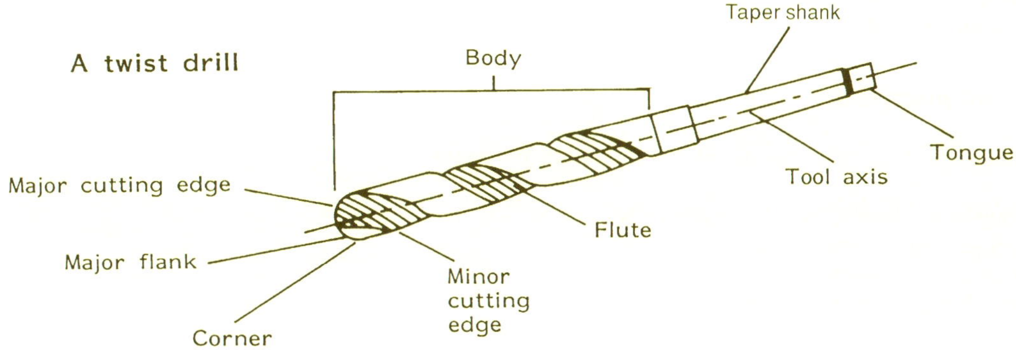


The basic terminology for multi-point tool 'teeth', e.g. the drill illustrated, is the same as that for a single-point tool. Some typical cutting applications for multi-point tools are shown in the next figure.

Multi-point tool cutting actions

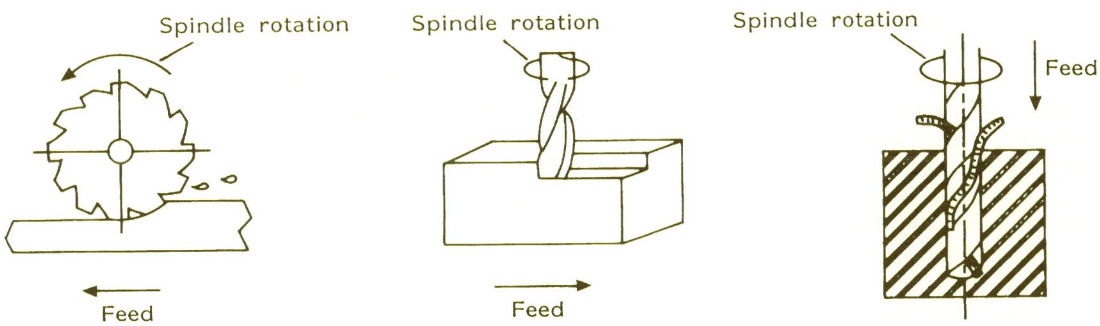


Horizontal milling　　Vertical milling　　Drilling

### 8.1.4 Cutting feed and cutting speed

The two main parameters associated with a cutting tool action and the corresponding workpiece are *cutting feed* and *cutting speed.*

The *cutting feed* is the rate at which the tool is fed into the workpiece or the rate at which the workpiece is fed into or past the tool. This is expressed in mm/rev, mm/cycle or mm/minute depending on the machining process being used.

The *cutting speed* is the rate at which the cutting edge passes over the work, expressed in m/minute.

Recommended cutting feeds and speeds for different types of machining operations and materials are given in the Annex at the end of this section.

### 8.1.5 Accuracy of tolerance and surface finish

The accuracy of tolerance and surface finish obtained during any machining process using a cutting tool will depend on the following factors:

(a) tool geometry and accuracy
(b) correct cutting speeds and feeds
(c) the rigidity and age of the machine tool
(d) the rigidity and accuracy of the tool and the component set-up
(e) machinability of the component material
(f) the coolant used.

### 8.1.6 Tool materials

Tool materials must exhibit some of the following properties:

(a) stronger and harder than the material to be cut
(b) tough enough to withstand shock loads
(c) retain a keen cutting edge
(d) retain hardness at high temperatures.

No tool material covers all these criteria, therefore a number of different types are available for different applications.

- *High carbon steel* Fully hardened and suitably tempered, retains cutting edge up to 225°C.
- *High-speed steel* Used for all standard tools. Fully heat treated. Alloy steel with main alloying element tungsten. Retains the cutting edge up to a temperature of 660°C.
- *Carbide* Used to form a clamped or screwed tip which is attached to a carbon steel or high speed steel body. Produced by powdered metallurgy, they are made from powdered carbides of tungsten, titanium, etc. mixed with nickel or cobalt, compressed at high temperature in a mould of the required shape. The cutting speed obtained is 3–6 times that of high-speed tools and therefore much greater metal removal rates are possible. Retains the cutting edge to temperatures in excess of 1000°C.

While carrying out a machining operation on a component, whether it be turning, milling, drilling, etc., a coolant must be used where required to improve the cutting action of the tool. The coolant is applied directly at the point where the cutting tool meets the workpiece.

The main functions of the coolant are:

(a) to reduce friction between chip, tool and workpiece
(b) to cool both the tool and work
(c) to wash away swarf
(d) to improve surface finish
(e) to increase tool life
(f) to help prevent chip welding.

### 8.1.7 Chips

Liquid coolants for materials such as mild steel and aluminium are usually in the form of synthetic oil or soluble oil, which is a mixture of oil and water. However, some materials require no coolant, e.g. cast iron, brass.

Depending on factors such as rake angle, cutting fluid, material, tool condition, feed and cutting speed, different types of chips may be formed at the point where cutting takes place.

The *discontinuous chip* is obtained when machining brittle materials such as brass and cast iron. High cutting forces induce tool wear. If such a chip is obtained while cutting soft materials, it implies that the rake angle is too low or the cutting feed is too high.

A *simple continuous* chip is most desirable for steels, aluminiums and other ductile materials. A long chip is usually obtained along with a lower chip speed, which gives more chance of obtaining a built-up edge.

Types of chip

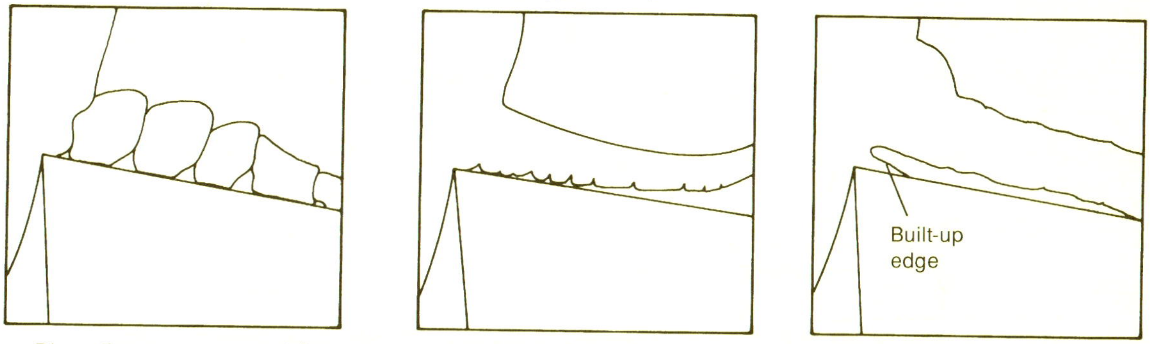


Discontinuous or segmental — Simple continuous — Continuous with built-up edge

A *built-up edge* is produced on the tip of a tool during machining. Due to high temperatures small pieces of material weld onto the tool tip. In some cases a small built-up edge can be beneficial in that it can insulate the tool tip from high temperatures. However, if it gets too large it can adversely affect the cutting edge causing a poor finish and tool breakage.

As shown in the figure, 70 per cent of the heat generated during the cutting process goes to the chip. Therefore increasing chip speed minimizes heat flow to the tool by reducing chip/tool contact time. The factors affecting good chip formation are also shown in the table.

Tool temperature* distribution during cutting (no coolant)

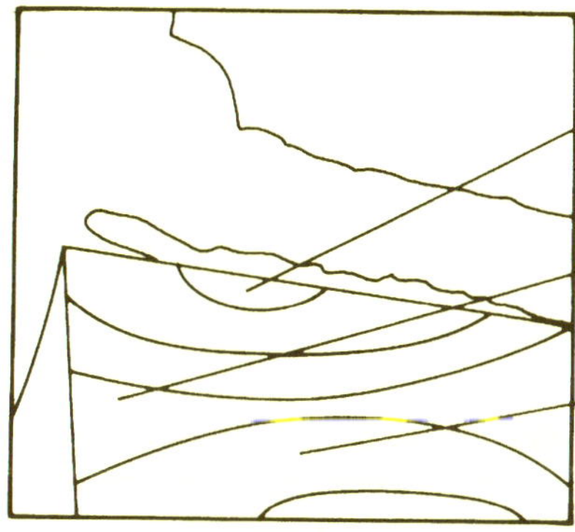


Heat distribution:

15% Workpiece
15% Tool
70% Chip

*Note that the temperatures shown should be regarded as indicating relative levels because the workpiece was pre-heated to allow infrared temperature detection.

*Factors in chip formation*

| *Detail* | *Simple continuous* | *Continuous with built-up edge* |
|---|---|---|
| Rake angle | Sufficient to give chip flow | Insufficient |
| Friction | Low | High |
| Cutting speed | Correct | Incorrect – usually too low |
| Tool state | Sharp | Blunt |
| Feed | Correct | Too high |
| Cutting fluid | Efficient | Inefficient |

### 8.1.8 Approximate machining time calculations for drilling, turning and milling

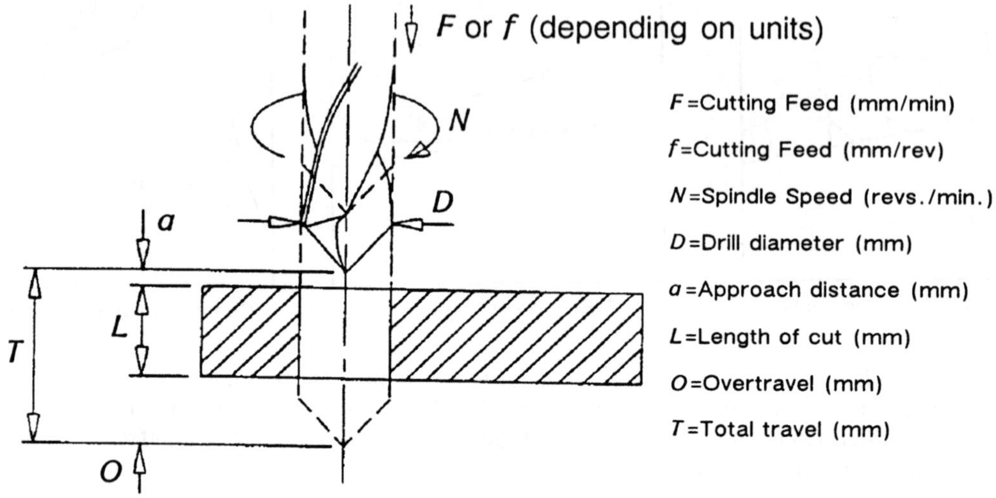


DRILLING SCHEMATIC

#### 8.1.8.1 Drilling

Drilling machining times can be calculated as follows.

The general formula for cutting speed is:

$$S = \frac{\pi DN}{1000}$$

where:

$S$ = cutting speed (m/min)
$D$ = diameter of drill (mm)
$N$ = revolutions/minute (revs/min)

In practice it is usually the revs/minute that are required to be calculated in order to set the machine to the proper spindle speed, the cutting speed being pre-determined by the material and the tool type (see the tables at the end of this section), therefore:

$$N = \frac{1000S}{\pi D}$$

**Example**

Find the revs/minute suitable for a 10 mm drill to cut free-cutting mild steel at 30 m/min.

$$N = \frac{1000S}{\pi D} = \frac{1000 \times 30}{\pi \times 10} \approx 955 \text{ revs/min}$$

The machining time ($t$) for drilling a component may also be calculated once the feed rate (mm/rev) and revs/min of the spindle are known.

Note that the overtravel ($O$) is the allowance for the point at the end of the drill. This may vary according to drill diameter. Typical drill point included angle is 118°. An approximate rule to calculate the depth of the point is to use the following to one decimal place:

$$O = \frac{1}{3} D$$

## Example

Calculate the machining time for drilling a 15 mm diameter through hole to a depth of 25 mm assuming:

$f$ = 0.25 mm/rev
$S$ = 30 m/min
$a$ = 3 mm
$L$ = 25 mm
$D$ = 15 mm
$t$ = total machining time (mins)

$$O = \frac{D}{3} = 5\text{ mm}$$

$$N = \frac{1000S}{\pi D} = \frac{1000 \times 30}{\pi \times 15} = 636.62\text{ revs/min, say } 630\text{ revs/min}$$

*Note:* The spindle speed is usually rounded down the the next available machine speed setting.

$$T = L + a + O = 25 + 3 + 5 = 33\text{ mm}$$

Therefore:

$$t = \frac{T}{F} = \frac{T}{fN} = \frac{33}{0.25 \times 630} = 0.21\text{ minutes}$$

Calculations of actual manufacturing operation times should also include allowances for set-up times, tool changing and non-machining feed movements. Note that the value calculated in this example only relates to machining time.

NB For blind holes calculate the point allowance. For feeds in between drill sizes interpolate as required.

### *8.1.8.2 Turning*

Turning machining times can be calculated as follows. The schematic diagrams outline all the variables for facing and plain turning.

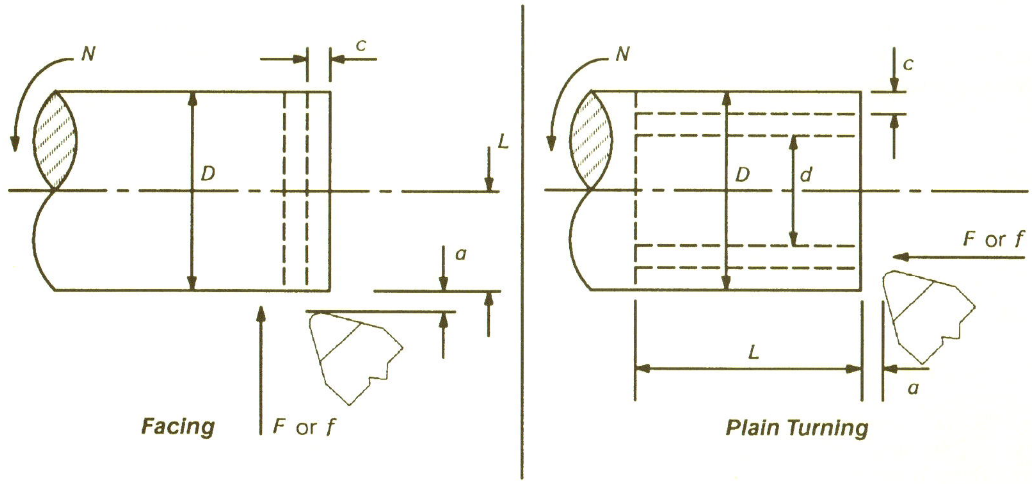


TURNING SCHEMATIC DIAGRAMS

$N$ = Spindle speed (revs./min.);
$C$ = Depth of cut (mm)
$F$ = Tool feed (mm/min)
$D$ = Initial cutting diameter (mm)
$a$ = Approach Distance (mm)
$f$ = Tool feed (mm/rev)
$d$ = Final machined diameter (mm)
$L$ = Length of cut (mm)

To calculate the cutting speed of the work in relation to the tool, the circumference of the workpiece is multiplied by the spindle speed in revs/min. That is:

$$S = \frac{\pi DN}{1000}$$

where:

$D$ = the diameter of the workpiece (mm)
$N$ = spindle speed (revs/min)
$S$ = cutting speed m/min.

### Example

Calculate the cutting speed of a 40 mm diameter bar at 140 revs/min.

$$S = \frac{\pi DN}{1000} = \frac{\pi \times 40 \times 140}{1000} = 18 \text{ m/min.}$$

As for drilling, the cutting speed is usually already known, therefore the spindle speed needs to be calculated and set.

### Example

Find the spindle speed suitable for turning a 40 mm diameter mild steel bar at 25 m/min.

From above:

$$N = \frac{1000S}{\pi D}$$

Therefore:

$$N = \frac{1000S}{\pi D} = \frac{1000 \times 25}{\pi \times 40} = 199 \text{ revs/min.}$$

For some turning operations *roughing* tools are used to remove large amounts of material and the surface finished with a *finishing* tool. The depth of cut for these operations is given in the Annex along with turning feeds and speeds.

### Example

Calculate the machining time for required for the following operations:

(a) facing a 40 mm diameter mild steel bar in one cut
(b) turning a 60 mm length on a 40 mm diameter mild steel bar in two cuts to a finish diameter of 30 mm.

Assume the following:

$f$ = 0.25 mm/rev
$S$ = 25m/min
$a$ = 2mm
$D$ = diameter of bar (mm)
$L$ = length of cut (mm)
$N$ = revs/min.
$T$ = total travel (mm)
$nc$ = number of cuts
$t$ = machining time (mins)

(a) $T = L + a = 20 + 2 = 22\,\text{mm}$

$$N = \frac{1000S}{\pi D} = \frac{1000 \times 25}{\pi \times 40} = 199\,\text{revs/min}$$

Therefore:

$$t = \frac{Tnc}{F} = \frac{Tnc}{fN} = \frac{22 \times 1}{0.25 \times 199} = 0.44\text{ mins}$$

(b) $N = 199$ revs/min as before

$$T = L + a = 60 + 2 = 62\,\text{mm}$$

Therefore:

$$t = \frac{Tnc}{F} = \frac{Tnc}{fN} = \frac{62 \times 2}{0.25 \times 199} = 2.48\text{ mins}$$

*Note:* Theoretically after every cut along the length, a new value of $N$ should be calculated. This is not normally done in practice, unless a very large reduction in diameter takes place. Also, different feeds and speeds may be used for roughing and finishing cuts.

If the number of cuts for plain turning are to be calculated then:

$$nc = \frac{(D - d)}{2c}$$

This number is rounded down for the number of roughing cuts; the material left is the depth of cut for the single finishing cut.

As with drilling, actual manufacturing time calculations should also include set-up times and non-machining feed moves.

Usually the machine you are calculating times for has a fixed speed range. It is likely that a speed of 199 revs/min will not be available, therefore the next speed down in the range is usually taken.

### *8.1.8.3 Milling*

Milling (in this case vertical) machining times can be calculated as follows. The schematic diagram outlines all the variables. Usually the overtravel is not as great but for inexperienced operators it is a safer option.

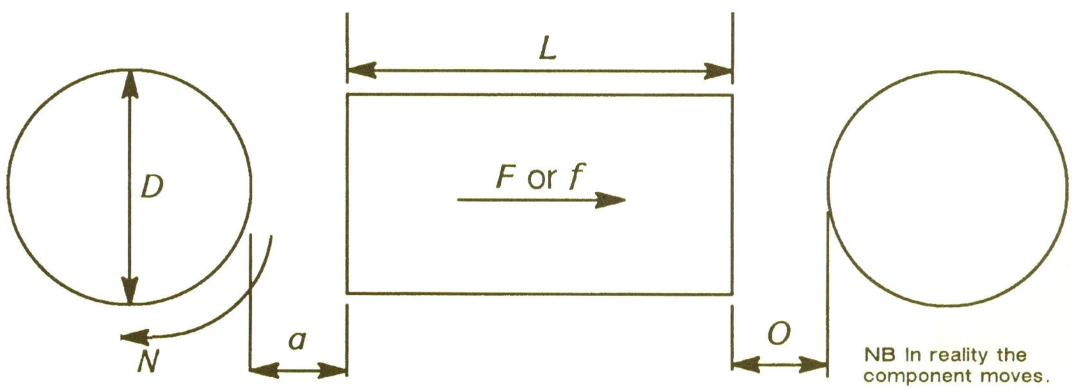


MILLING SCHEMATIC (Plan view)

$N$ = Spindle speed (revs./min.); $D$ = Cutter diameter (mm) $a$ = Approach Distance (mm)
$L$ = Length of cut (mm) $O$ = Overtravel (mm) $F$ = Tool feed (mm/min)
$f$ = Tool feed (mm/rev)

Assume the maximum depth of cut for face mills is usually 12 mm for mild steel. For end mills and slot drills it is is usually *half the cutter diameter.*

Cutting speeds for milling are usually expressed in m/min. That is:

$$S = \frac{\pi DN}{1000}$$

where:

$S$ = cutting speed (m/min)
$D$ = diameter of cutter (mm)
$N$ = spindle speed (revs/min)

However, as before it is usually the spindle revolutions which need to be determined:

$$N = \frac{1000S}{\pi D}$$

Again, this is usually rounded down to the next available machine speed.

### Example

Find the appropriate spindle speed for milling a mild steel component using a 100 mm diameter plain milling cutter at a cutting speed of 25 m/min.

$$N = \frac{1000 \times 25}{\pi \times 100} = 80\,\text{revs/min}$$

The tabulated *feed rate* for milling is usually expressed in mm/tooth. The *feed rate* is the rate at which the work advances past the cutter (mm/rev). That is:

Feed rate = Feed per tooth × number of teeth (mm/rev)

Also:

Feed rate in mm/min = Feed per tooth × no. of teeth × $N$

The feed rate per tooth for the same material differs depending on the type of cutter being used. Tables outlining facing milling feeds and cutting speeds are given in the Annex the end of this section.

Knowing the feed, cutting speed and cutting length it is possible to calculate the machining time.

### Example

Calculate the machining time for face milling an 80 mm long, 40 mm wide block of mild steel in one cut with a 15 tooth, 50 mm diameter cutter. Assume:

| | |
|---|---|
| Cutter approach $a$ = 5 mm | Overtravel $O$ = 5 mm |
| Feed rate $f$ = 0.30 mm/tooth | Cutting speed $S$ = 25 m/min |
| Feed rate $F$ = mm/min | $N$ = revs/min |
| $D$ = cutter diameter (mm) | $n$ = no. of teeth |
| $L$ = length of workpiece (mm) | $t$ = cutting time (mins) |
| $T$ = total travel (mm) | $nc$ = 1 |

$$N = \frac{1000S}{\pi D} = \frac{1000 \times 25}{\pi \times 50} \approx 160\,\text{revs/min}$$

$$F = nfN = 15 \times 0.30 \times 160 = 720\,\text{mm/min}$$

$$T = a + O + L + D = 5 + 5 + 80 + 50 = 140\,\text{mm}$$

$$t = \frac{Tnc}{F} = \frac{140 \times 1}{720} = 0.19\,\text{mins}$$

Although not all machining operations and processes are covered in this section, it will usually be possible to develop formulae for other machining times by using the principles outlined here.

### 8.1.9 Annex: EA project drilling, turning and milling speeds and feeds

*Drilling feeds and speeds (High-speed steel)*

| Material | Cutting speed (m/min) | Drill diameter (mm) | | | |
|---|---|---|---|---|---|
| | | 6 | 12 | 18 | 25 |
| | | feed (mm/rev) | | | |
| Aluminium | 195–390 | 0.30 | 0.40 | 0.61 | 0.81 |
| Brass | 60–75 | 0.20 | 0.25 | 0.41 | 0.51 |
| Cast iron | 25–35 | 0.13 | 0.18 | 0.25 | 0.31 |
| Mild steel | 30–36 | 0.20 | 0.25 | 0.41 | 0.51 |
| Stainless steel | 6–15 | 0.13 | 0.18 | 0.25 | 0.31 |
| Plastics | 20–75 | 0.10 | 0.15 | 0.20 | 0.25 |

*Turning feeds and speeds (High-speed steel)*

| Material | Feed (mm/rev) | Cutting speed (m/min) |
|---|---|---|
| Aluminium | 0.1 | 60–150 |
| Brass | 0.38 | 45–120 |
| Cast iron | 0.25 | 22–30 |
| Mild steel | 0.25 | 25–45 |
| Stainless steel | 0.38 | 9–18 |
| Plastics | 0.1 | 9–120 |

*Note*: Cutting speeds can be increased 3–6 times for carbide tipped tooling. Recommended depths of cut are as follows 2–5 mm for roughing depending on material type; 0.3–0.5 mm for finishing.

*Milling feeds and speeds (High-speed steel)*

| Material | Feed (mm/tooth) | Cutting speed (m/min) |
|---|---|---|
| Aluminium | 0.13–0.76 | 60–150 |
| Brass | 0.25–0.76 | 60–150 |
| Cast iron | 0.13–0.63 | 15–30 |
| Mild steel | 0.25–0.63 | 25–35 |
| Stainless steel | 0.10–0.38 | 9–20 |
| Plastics | 0.25–0.80 | 20–160 |

*Note*: Decrease speeds by 40% for side and face milling or end milling and 60% for form milling. Recommended depths of cut are a maximum of half the cutter diameter for roughing and 0.1–1.6 mm for finishing, except for facing.

Note that the speeds and feeds quoted here are for use in the projects in the main text. Consequently, these are lower than those normally used in industry. However, they still act as a useful guide for machining calculations. Normally the tooling used in industry tends to be carbide or some other exotic material. To obtain further information on the speeds and feeds for these types of tooling refer to *Newnes Mechanical Engineer's Pocket Book* by Roger Timings and Tony May, second edition (1997).

## 8.2 Numerical control (NC) machining

### *8.2.1 Terminology and formats*

Although the letters and formats used in NC programming have been largely standardized, variations still occur between controllers from different manufacturers. Thus the information given in this section, although accurate, probably cannot be applied to all controllers. However, it does give a good general outline of programme structures and associated programming codes.

### *8.2.2 Block numbers – 'N' (N1–N9999)*

Each line of coded information in an NC programme is called a block and is given a reference number for ease of identification both by the programmer and the machine controller. Although the controller will allow a random sequence of block numbers to be used and will search through the whole program for a nominated block, ascending sequences – either 1, 2, 3, etc. or 10, 20, 30, etc. – are generally used. Programming in tens allows additional moves to be inserted later if the original program proves to be over-optimistic or if information has been inadvertently omitted.

### *8.2.3 Preparatory functions – 'G' codes*

The most commonly used 'G' codes are listed below. Those marked 'M' are modal; i.e. once called up they remain effective until subsequently cancelled later in the program either by a direct countermand (e.g. G40 to cancel G41 or G42) or by a revised command (e.g. G00, G01, G02 and G03 mutually cancel or replace each other):

| | | |
|---|---|---|
| **G00** | **'M'** | Rapid positioning, point to point (M) |
| **G01** | **'M'** | Linear interpolation at controlled feed rate |
| **G02** | **'M'** | Clockwise circular interpolation |
| **G03** | **'M'** | Counter-clockwise circular interpolation |
| **G04** | | Dwell |
| **G17** | | Plane selection *X*/*Y* |
| **G40** | | Cancel cutter diameter compensation |
| **G41** | **'M'** | Cutter diameter compensation left |
| **G42** | **'M'** | Cutter diameter compensation right |
| **G50** | | Define datum point |
| **G53** | | Cancel set-up offsets |
| **G70** | **'M'** | All dimensions in inches |
| **G71** | **'M'** | All dimensions in millimetres |
| **G80** | | Cancel canned cycles (for list of canned cycles see next section) |
| **G90** | **'M'** | All X, Y and Z dimensions absolute |
| **G91** | **'M'** | All X, Y and Z dimensions incremental |
| **G92** | | Define datum point |
| **G94** | **'M'** | Feed rate per minute |
| **G95** | **'M'** | Feed rate per revolution |
| **G96** | **'M'** | Constant surface speed |
| **G97** | **'M'** | Speed in revolutions per minute. |

### 8.2.4 Canned cycles

Canned cycles combine a sequence of moves so that they can can be called up by a single G code.
Machining centre canned cycles usually include:

| | | |
|---|---|---|
| **G80** | **'M'** | Cancels canned cycles |
| **G81** | **'M'** | Drilling |
| **G82** | **'M'** | Drilling with dwell |
| **G83** | **'M'** | Deep hole drilling with pecks |
| **G84** | **'M'** | Tapping |
| **G85** | **'M'** | Boring. |

These can be represented schematically relative to a reference plane defined in the program, i.e. the feed move starts at the height defined by the reference plane and continues to the depth defined by the Z dimension programmed.
Turning centre canned cycles usually include:

| | | |
|---|---|---|
| **G66** | **'M'** | Contouring |
| **G67** | **'M'** | Contouring |
| **G68** | **'M'** | Roughing |
| **G69** | **'M'** | Roughing |
| **G81** | **'M'** | Turning |
| **G82** | **'M'** | Facing |
| **G83** | **'M'** | Deep hole drilling |
| **G84** | **'M'** | Threading |
| **G85** | **'M'** | Threading |
| **G88** | **'M'** | Grooving. |

### 8.2.5 Dimensions 'X', 'Y', 'Z'

With absolute programming, each point within the reach of the machine is defined in three-dimensional space by its X, Y and Z coordinates relative to a pre-defined datum (0, 0, 0). The Z axis is always parallel to the spindle axis. For machining centre work, Z0 is usually chosen as the top surface of the workpiece so cutting moves in Z are made towards negative (Z–) positions. X and Y may be positive or negative, depending on the choice of the (0, 0, 0) position relative to the workpiece.

In incremental programming (usually called up by G91), X, Y or Z moves (positive or negative) in the same block are relative to the current position of the cutter as established by its previous move.

### 8.2.6 Dimensions 'I', 'J' and 'K'

Circular moves (G02 or G03) are defined by their start/end points (X, Y, Z) and by the centre of the particular arc to be cut. I, J and K define the centre: with some controls I, J and K are absolute, while for others they are incremental. Some controls offer radius programming (with the letter R) as a simpler alternative.

### 8.2.7 Feed rates – 'F'

As in conventional milling and turning, feed rates are usually expressed in mm/min and mm/rev for machining and turning centres, respectively. On occasion it is desirable to change from one to another (by using G94 or G95) either for a whole program or, perhaps, more likely, for some small part of the program. Note that G94 and G95 are modal and cancel each other.

### 8.2.8 Spindle speed – 'S'

With machining centres, spindle speeds are almost always defined in revolutions/minute (rpm) but with turning work, where the diameter being cut is frequently or constantly changing (e.g. a stepped shaft or a facing operation, respectively), cutting speed is best defined in surface speed (m/min). In the program, the machine's control then calculates and commands the appropriate rpm at each instant.

### 8.2.9 Tool numbers and compensations – 'T', 'D'

The tool number T is usually the number of the position in the automatic tool changer magazine (or of the turret position in turning) where the tool is to be loaded. The tool length offset, the distance which the tool has to travel from the tool load/unload position for its tip to reach Z0, is recorded against a D number (usually the same as the T number) along with its cutter diameter compensation (for profiling). Thus, for example, T11D11 may refer to a 12 mm slot drill, loaded in the magazine in position 11, with a distance, say 234.56 mm, to travel before it reaches the top surface of the workpiece, recorded in D11 along with its diameter, 12 mm.

For external profiling the milling cutter should be programmed to move round the shape clockwise (for down-cutting and the best surface finish): the cutter will be travelling to the left of the surface so G41 will be used. For internal down-cutting the direction will be counter-clockwise but G41 will again be used. Cutter diameter compensation must be called up in a movement block before the blocks defining the profile: it is cancelled in a movement block after the profile has been completed.

### 8.2.10 Miscellaneous functions –'M' Codes

Apart from the preparatory G functions a number of other miscellaneous functions are required throughout any program, e.g. tool change, coolant on, spindle on. This usually takes the form of the letter M followed by two digits. The most common M codes are:

| | |
|---|---|
| **M00** | Program stop* |
| **M01** | Optional stop* |
| **M02** | End program* |
| **M03** | Spindle on – clockwise |
| **M04** | Spindle on – anti-clockwise |
| **M05** | Spindle off |
| **M06** | Tool change* |
| **M07** | Mist coolant on |
| **M08** | Flood coolant on |
| **M09** | Coolant off |
| **M30** | End of program – returns to start for next component.* |

* These commands operate only after all other commands in the block are completed. The other M codes remain operational through subsequent blocks until cancelled by a related M code.

## 8.3 Welding

### 8.3.1 Welding joint types and parameter definitions

**Gap** The distance between the components to be joined.

**Included angle** The angle between the welding faces after the parts to be joined have been positioned.

**Heat-affected zone** Those parts of the parent metal which have had their physical properties changed due to the high temperature of the welding (or cutting) process. This does not include melted material.

**Kerf** The space or gap left after a material has been thermally cut.

**Leg length** The distance from the apex of the angles formed by the two welding faces to the line where the weld face joins the parent metal.

**Parent metal** The material or component to be welded.

**Penetration** The distance that fusion goes beyond the base or root of the joint or the depth of the fusion zone in the parent metal.

**Reinforcement**:

(a) In a *butt joint*, the surplus metal deposited above the level required to make it flush
(b) Any excess metal deposited on the surface of the parent welding surface
(c) In a *fillet weld*, the metal deposited which is surplus to that required to give a mitre weld profile.

**Weld toe**:

(a) The line where the face of the weld joins the parent
(b) The line where the weld faces join the weld face of a previously created weld.

**Weld face** The surface of the weld seen from the side from which the weld was created.

**Weld junction** The fusion zone/heat-affected zone boundary.

**Weld metal** All metal which was melted during the course of the weld.

**Weld pool** The pool of liquid metal created or formed during fusion welding.

**Weld zone** The zone including the weld metal, the heat-affected zone and the fusion zone.

**Welding sequence** The order in which the various welding operations, joints and welds are to be carried out.

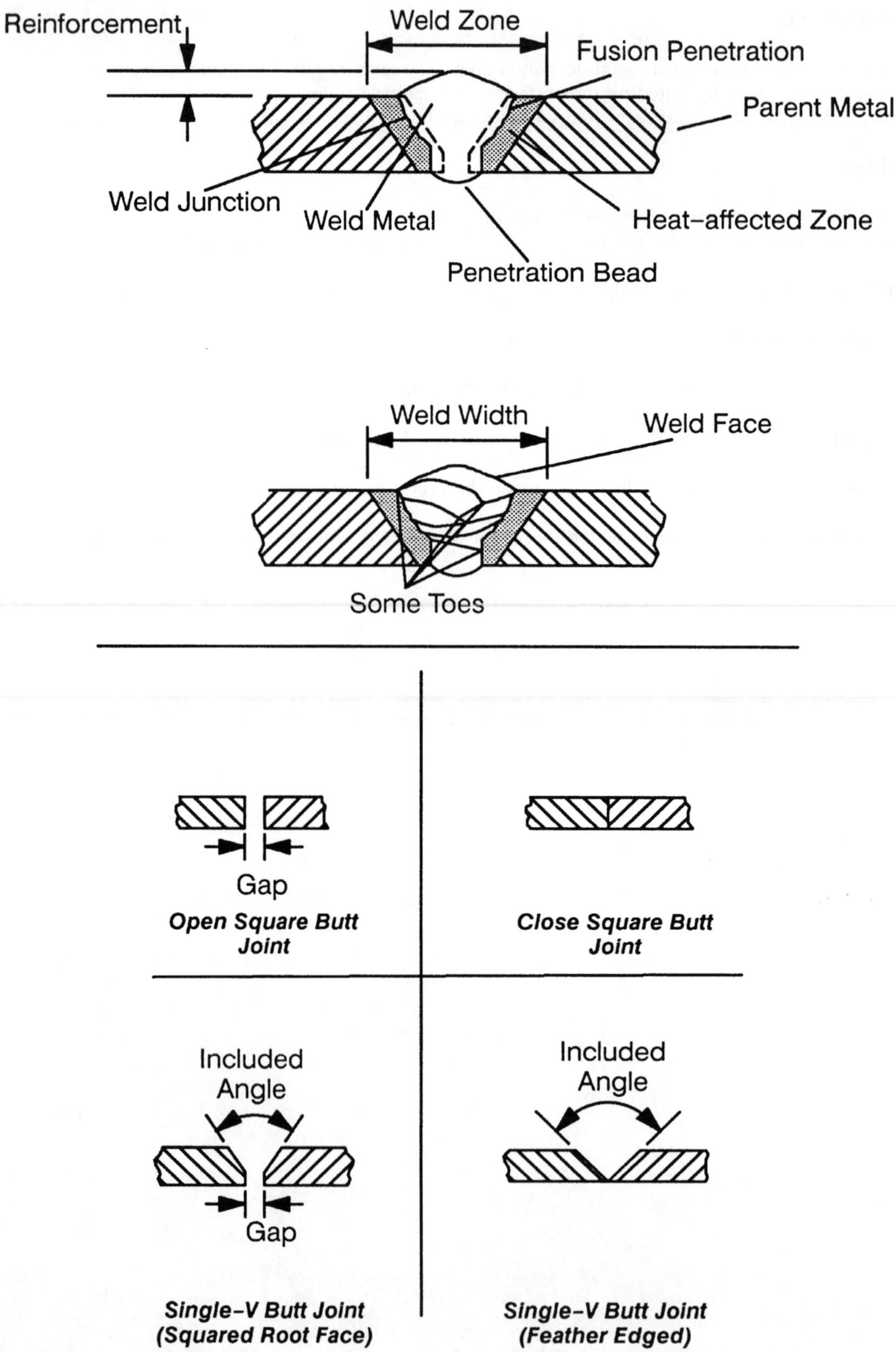


TYPICAL BUTT WELD PARAMETERS AND JOINTS

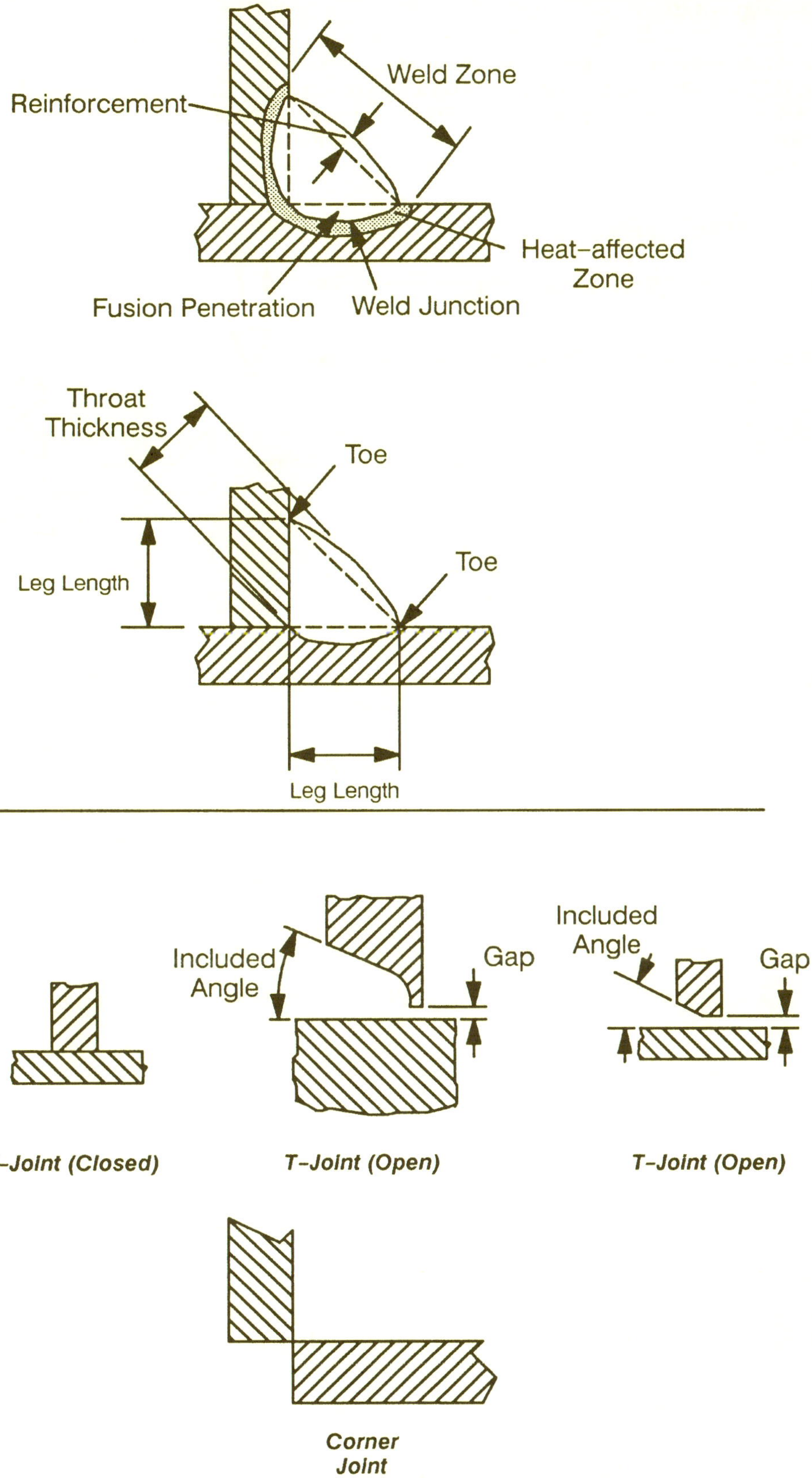


TYPICAL FILLET WELD PARAMETERS AND JOINTS

### 8.3.2 Drawing welds

Welds can be denoted on drawings by using the relevant national or international standard. In the United Kingdom, BS499 Section 7 defines the formal drawing requirements for welded components. Specific weld details, such as gaps, welding procedures or edge preparation, are usually covered by company-specific welding procedure or operation instruction sheets and not the BS499.

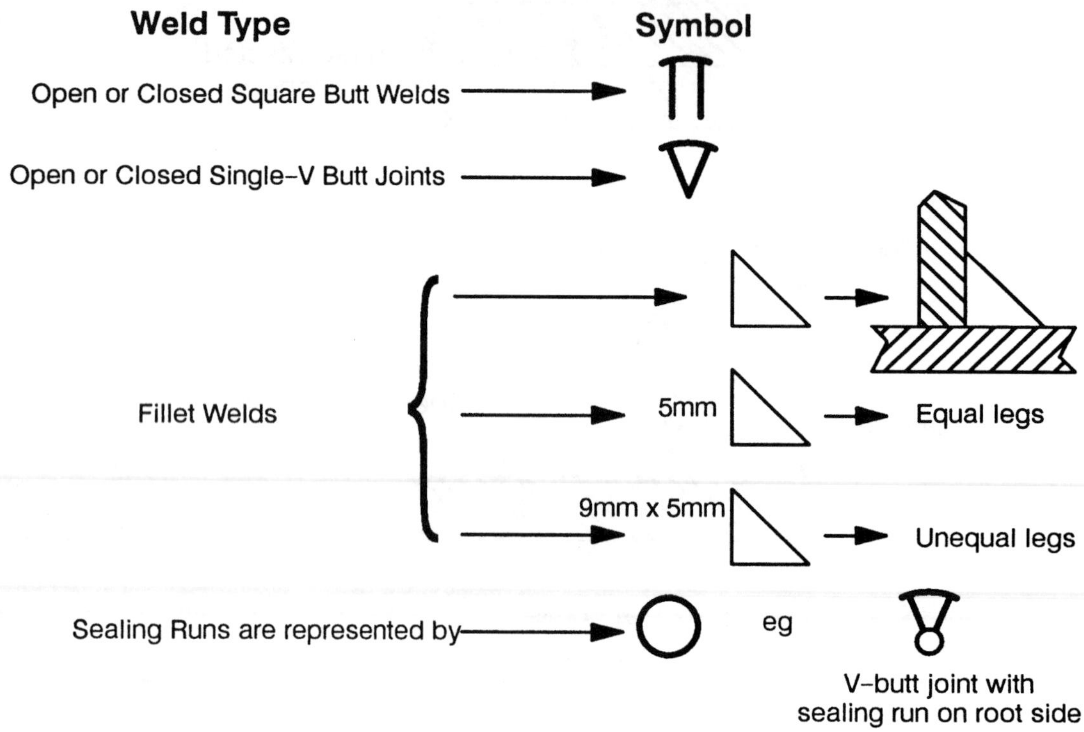


On drawings arrows show the position of the welded joint. The arrow is positioned on the side from which the joint is to be welded. Hence the descriptions shown in the diagrams below, namely 'arrow side' and 'other side'.

The basic requirements for dimensioning a weld on a drawing are as follows (see BS499 for more details):

(a) a joint symbol
(b) an arrow to represent the position of the weld
(c) the inclusion of weld detail information (if required)
(d) supplementary instructions.

*8.3.2.1 Regular intermittent or staggered intermittent fillet welds*

The 'arrow side' and 'other side' on welding drawings are defined as shown below, using BS499. A triangle below the line relates to the 'arrow side' and a triangle above the line relates to the 'other side'.

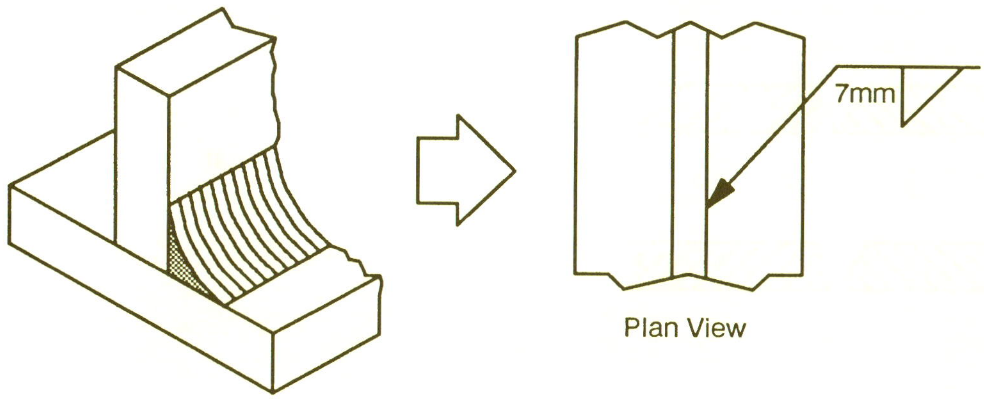


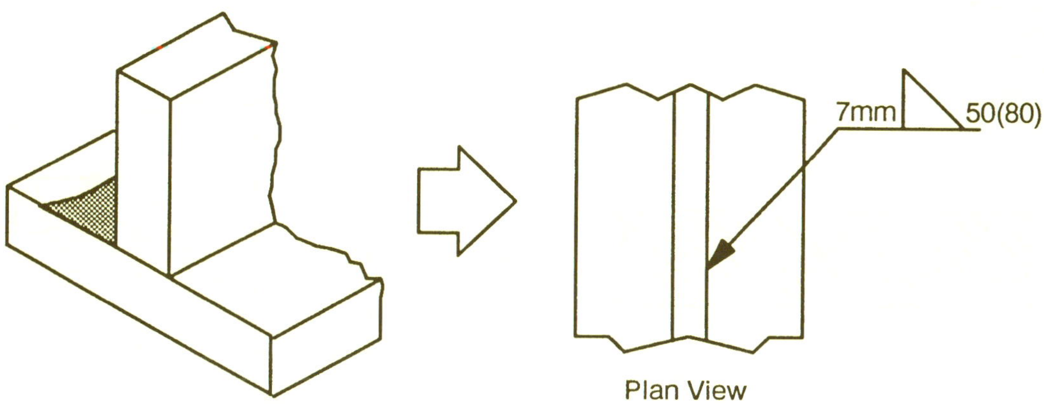


Some typical examples of weld drawing details are now given; these relate to gas welding and electric arc.

*8.3.2.2 Close square butt joint or open square butt joint*

### 8.3.2.3 Single-V butt joint with root edge or with root face

For a single-V butt joint with root edge, feather edge preparation is needed. For thicker materials a gap may also be required.

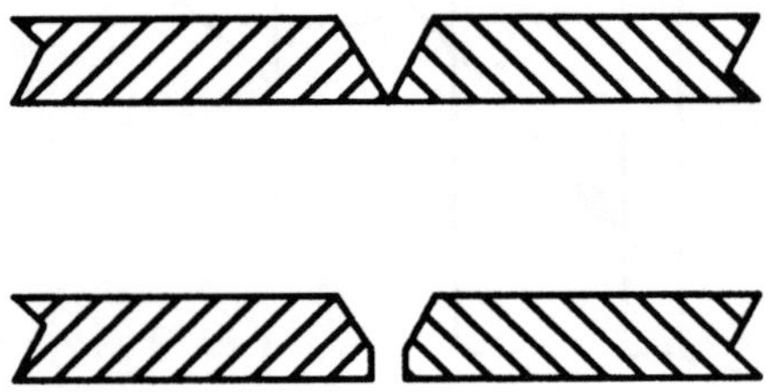

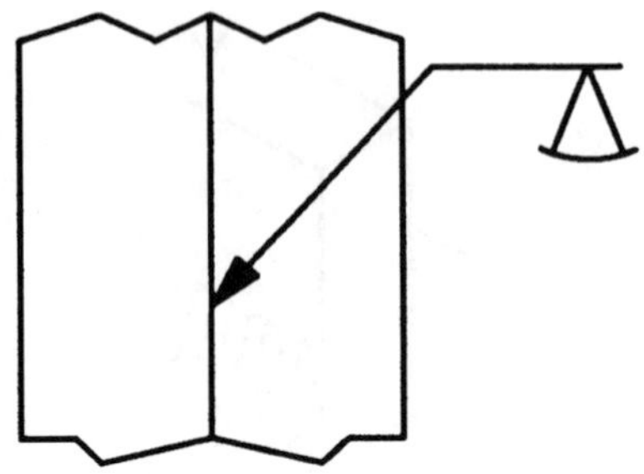

Plan View

### 8.3.2.4 Double-V butt joint

For a double-V butt joint sharp edge preparation is needed. For thicker materials a gap may also be required.

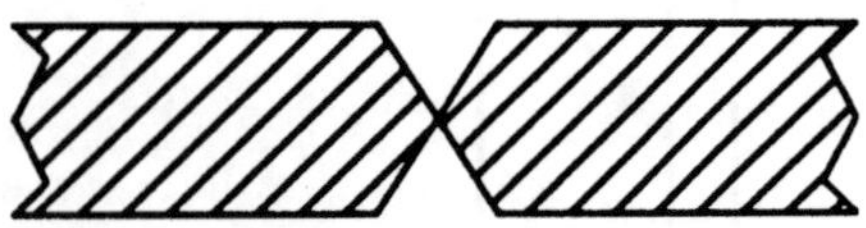

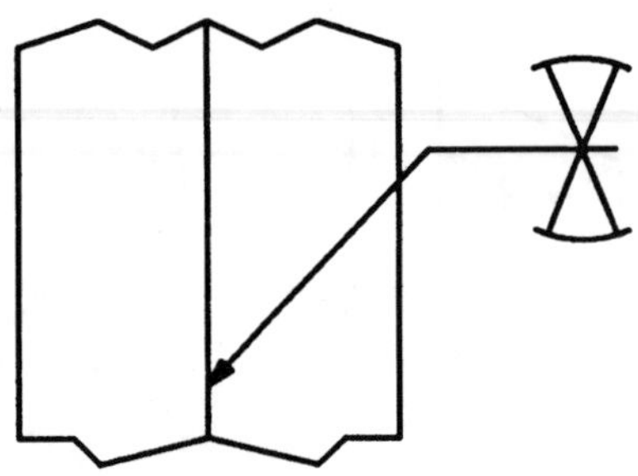

Plan View

### 8.3.2.5 Close square T-joint

The leg length of the weld is stated at the left-hand side of the symbol.

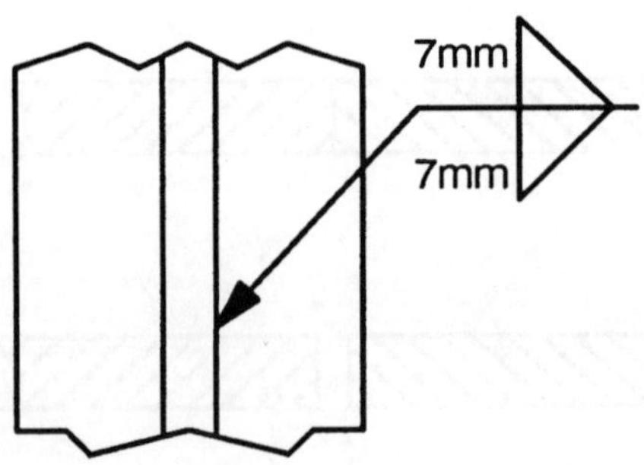


Plan View

### *8.3.2.6 Double lap joint*

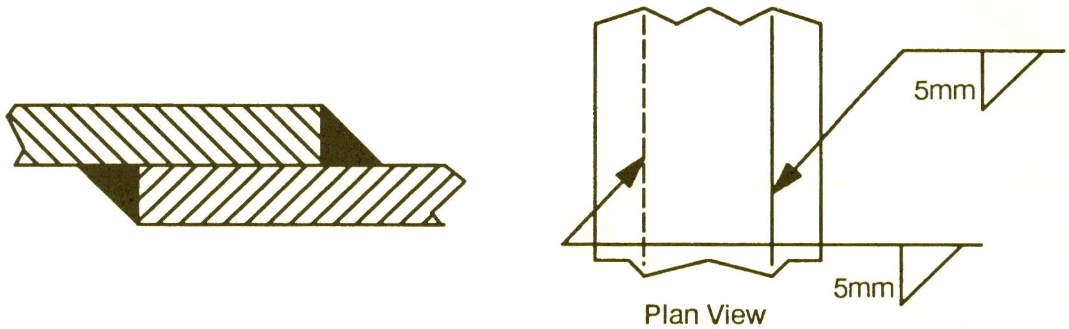


### *8.3.2.7 Close outside corner joint*

The leg length tends to be the same size as the material thickness unless otherwise indicated.

If the leg sizes are different on such a joint, then it is illustrated as shown below.

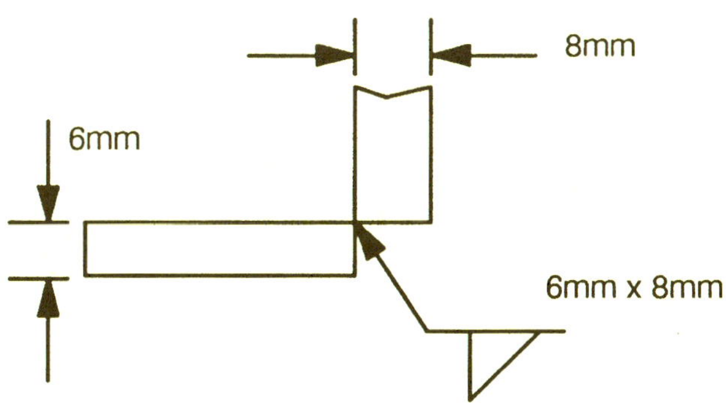

### *8.3.2.8 Radial closed inside corner joint*

For circular welds the drawing should illustrate the weld sizes and form as shown in the next figure. Note that the welds may or may not be represented on the drawing.

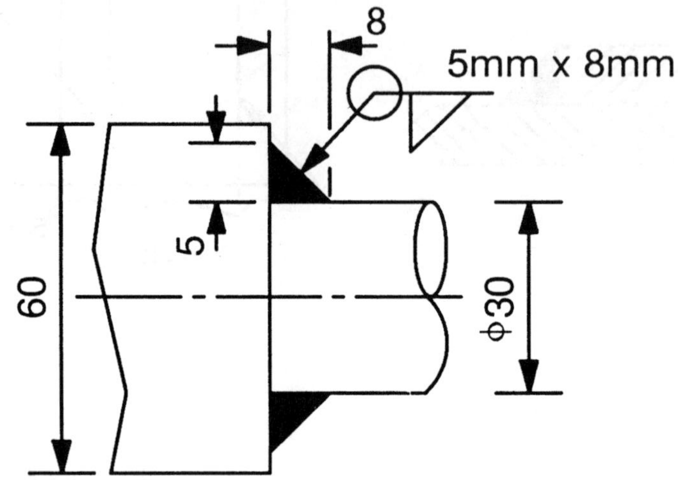


For a series of regular intermittent welds beginning at the edge on each side of a joint, the drawing should state the information as shown.

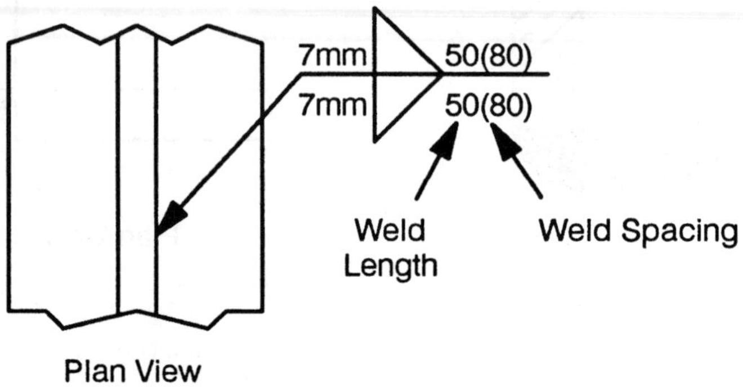


If the requirement is that the regular intermittent welds should start a specific distance from the edge, then the weld dimensioning format is amended as shown below.

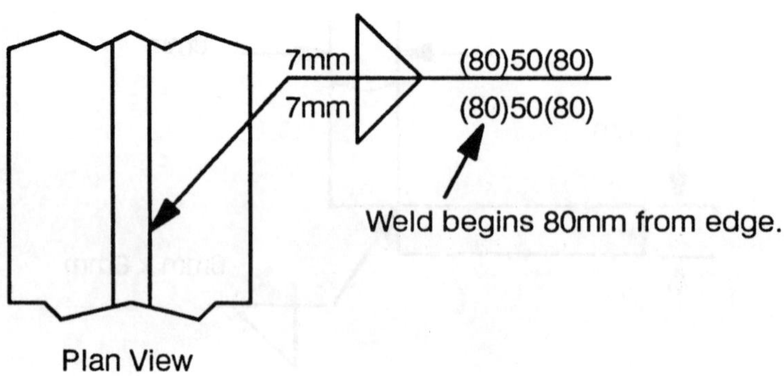

For staggered intermittent fillet welds an extra dimension is added to the end of the description outlining the next weld size. This can be further extended by more bracketed and non-bracketed values for additional welds.

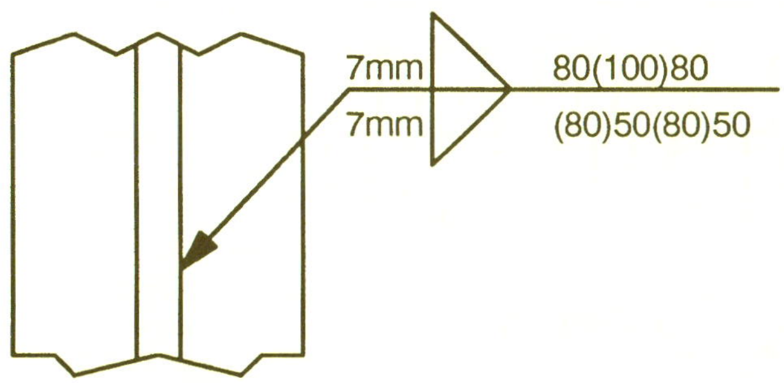


Plan View

In the above case the 'arrow side' welds start 80 mm from the edge, a weld of 50 mm in length is then produced, then a gap of 80 mm followed by a final weld of 50 mm and so on. For the 'other side' an 80 mm weld starting at the edge is produced, followed by a 100 mm gap then an 80 mm weld and so on.

## 8.4 Reading measuring instruments

### *8.4.1 The micrometer*

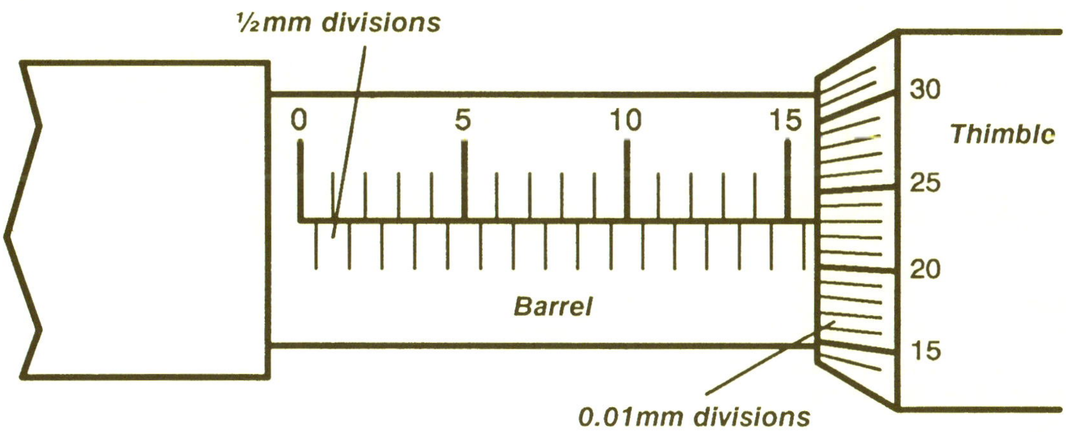


**Reading a metric micrometer**

| | *Method* | *Example in Figure* |
|---|---|---|
| Step 1 | Read the number of divisions above the barrel axial line | 15 = 15 mm |
| Step 2 | Read the number of additional ½ mm divisions below axial line | 1 = 0.5 mm |
| Step 3 | Read the value of the line on the thimble which coincides with the axial line | 23 = 0.23 mm |
| Step 4 | Add the values together | Total reading = 15.73 mm |

### 8.4.2 The vernier caliper gauge

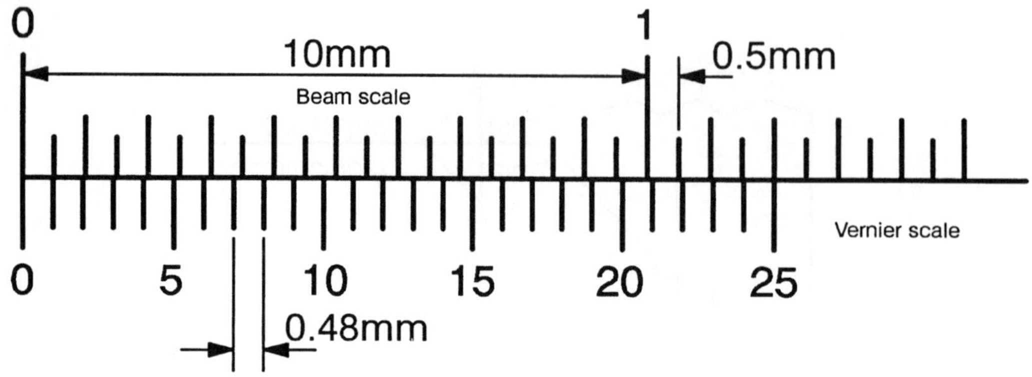


The *beam scale* is fixed on the main body support of the measuring instrument and is associated with the fixed jaw. The *vernier scale* is marked on a moving blade and is associated with the moving jaw.

On the beam scale the distance from '0' to '1' is 10 mm. This is further subdivided into 1 mm and 0.5 mm widths.

On the *vernier scale* a 12 mm length is subdivided into 25 equal parts; thus the width of each division is 0.48 mm. Therefore the difference between one small division on the beam scale and one small division on the vernier scale is 0.02 mm.

For example, as shown below, if the vernier scale is moved until the 10th mark is level with a mark on the beam scale then the movement is $10 \times 0.02 = 0.20$ mm.

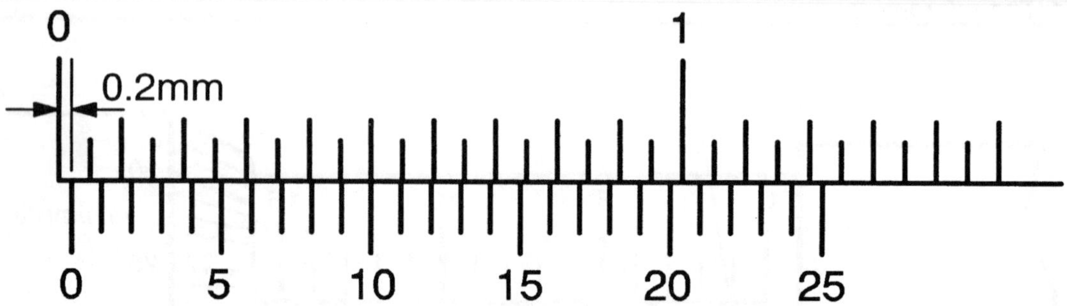


*Example*

When the jaws of the device are closed, both the left-hand zeros of the scales align, therefore the reading is zero. As the jaws are moved apart to lightly grip the part being measured then, once they are finally positioned, look for the point at which both scales exactly line up. Then carry out the procedure given below (see figure for example).

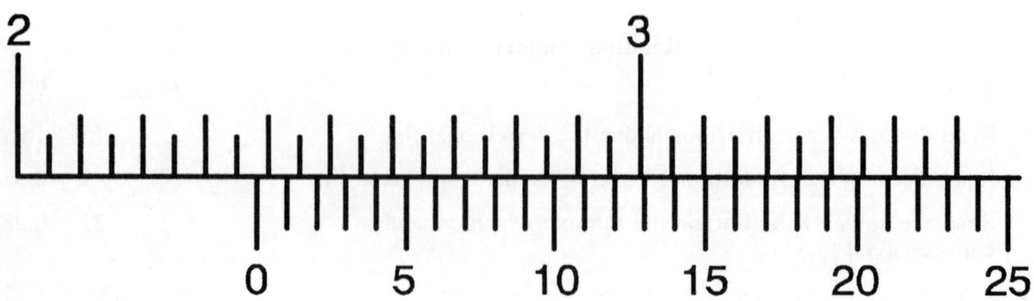

**Reading a metric vernier**

| | *Method* | *Example in Figure* |
|---|---|---|
| Step 1 | The distance between the zero on the beam scale and the zero on the vernier scale is 23.5 mm | 10 × 2.35 = 23.5 mm |
| Step 2 | Both scales match at mark 16 on the vernier scale | 16 × 0.02 = 0.32 mm |
| Step 3 | Add the values together | Total reading = 23.82 mm |

### *8.4.3 The vernier protractor*

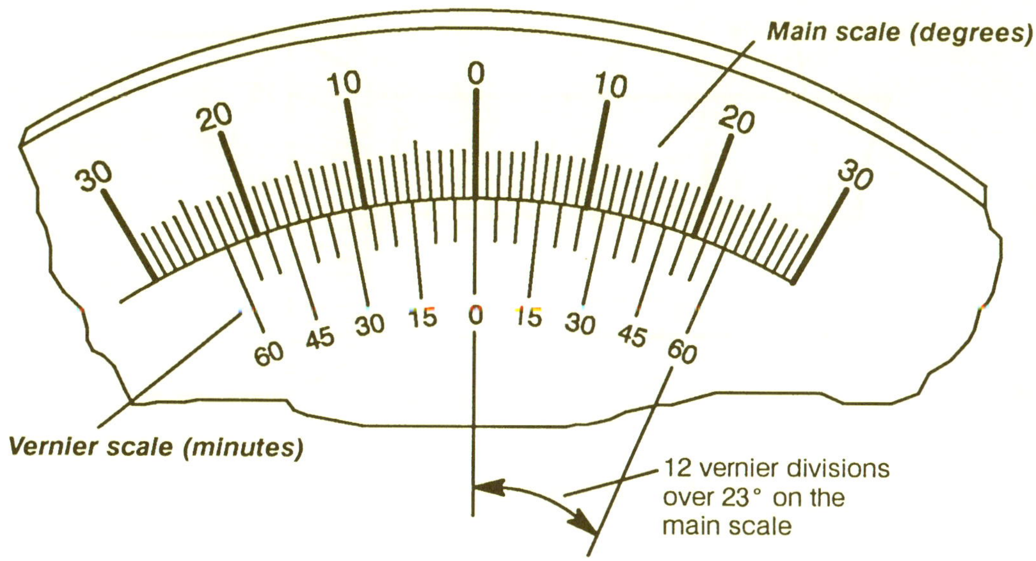


1 division on the vernier scale = 1 $^{11}/_{12}$° = 1° 55'.

Therefore, the basic protractor measurement is the difference between two divisions on the main scale and one division on the vernier, i.e. 5' (or $^{1}/_{12}$°).

*Example*

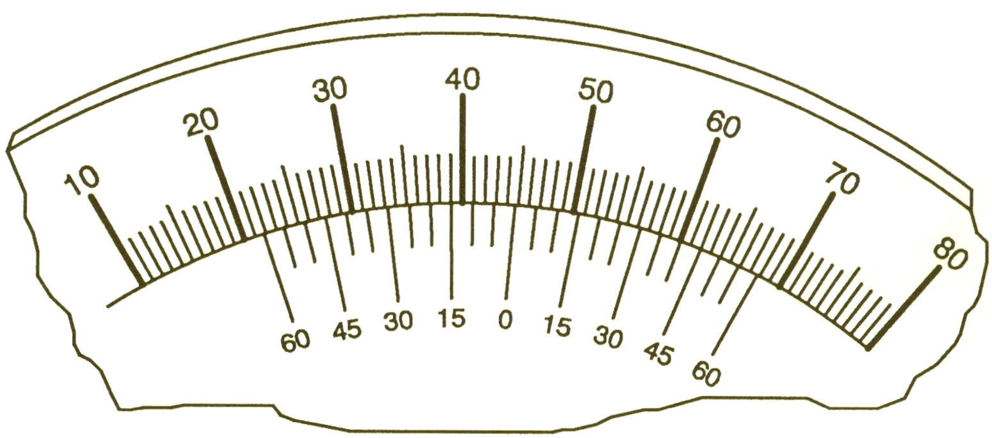

**Reading a vernier protractor**

| | *Method* | *Example in Figure* |
|---|---|---|
| Step 1 | Read the number of degrees between zero on the main scale and zero on the vernier | 44° |
| Step 2 | Read the number of minutes from the zero on the vernier scale to the point where a vernier line coincides with the main scale | 45' |
| Step 3 | Add the values together | Total reading (left to right) = 44° 45'<br>Total reading (right to left) = 135° 15' |

### 8.4.4 The depth gauge

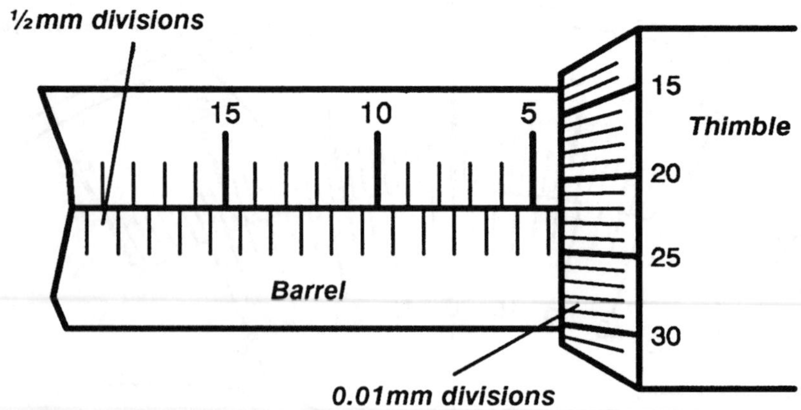


**Reading a metric depth gauge**

| | *Method* | *Example in Figure* |
|---|---|---|
| Step 1 | Read the number of full divisions above the barrel axial line | 4 = 4 mm |
| Step 2 | Read the number of additional ½ mm divisions below axial line | 0 = 0 mm |
| Step 3 | Read the value of the line on the thimble which coincides with the axial line | 22 = 0.22 mm |
| Step 4 | Add the values together | Total reading = 4.22 mm |

### 8.4.5 Slip gauges

Slip gauges, or 'slips' as they are more commonly called, are specially manufactured blocks of steel which have been hardened, ground and polished to a very high accuracy. The slips are wrung together to form a larger building block of the desired size for measuring, setting metrology instruments or calibration.

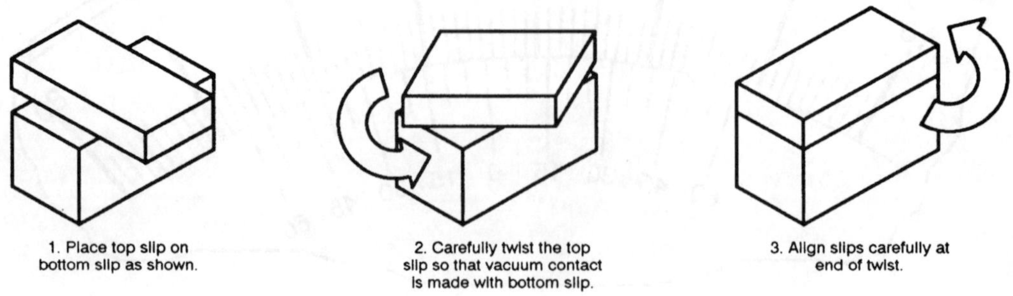

As they are wrung together, the slips form a small pile; their depths add up to the desired size. The minimum number of slips should be used where possible and they should be kept very clean as dirt between slip gauges can seriously affect the attachment process and the accuracy of the size, and cause slip wear.

A typical set up of slip gauges is shown here. Varying sizes of gauge are available; in this case sizes can be set to a value of 0.005 mm. The protector slips are used to protect the top and bottom of small piles of slips. When used their size (usually 2 mm each) must be taken into account. When worn and scratched they have to be replaced; however, this is more economical than replacing other slips which might be damaged if the protector slips are not used.

A typical example of the calculation for a pile of slips for a predetermined size is also given here.

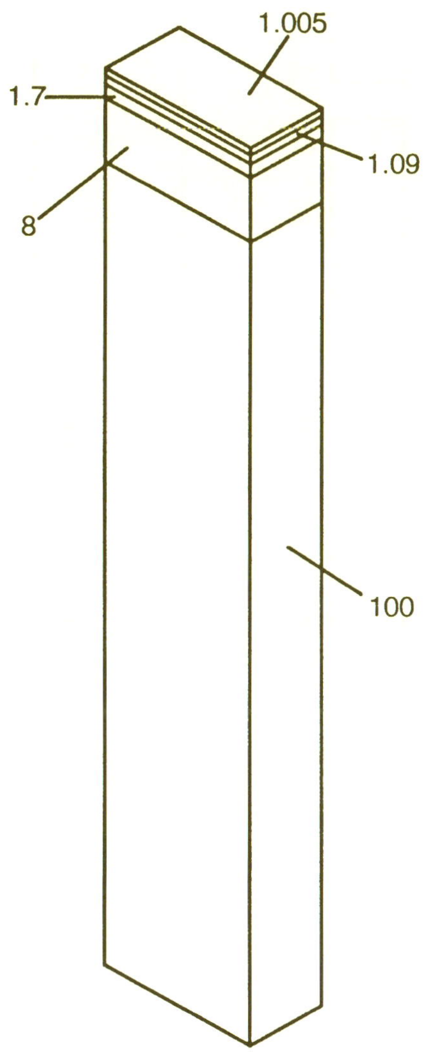


**Target Size - 111.795mm**

***111.795***
**– 1.005**

***110.790***
**– 1.090**

***109.70***
**– 1.700**

***108.000***
**– 8.000**

**100.000**

**Gauge Sizes**

(N.B. No protective slips used.)

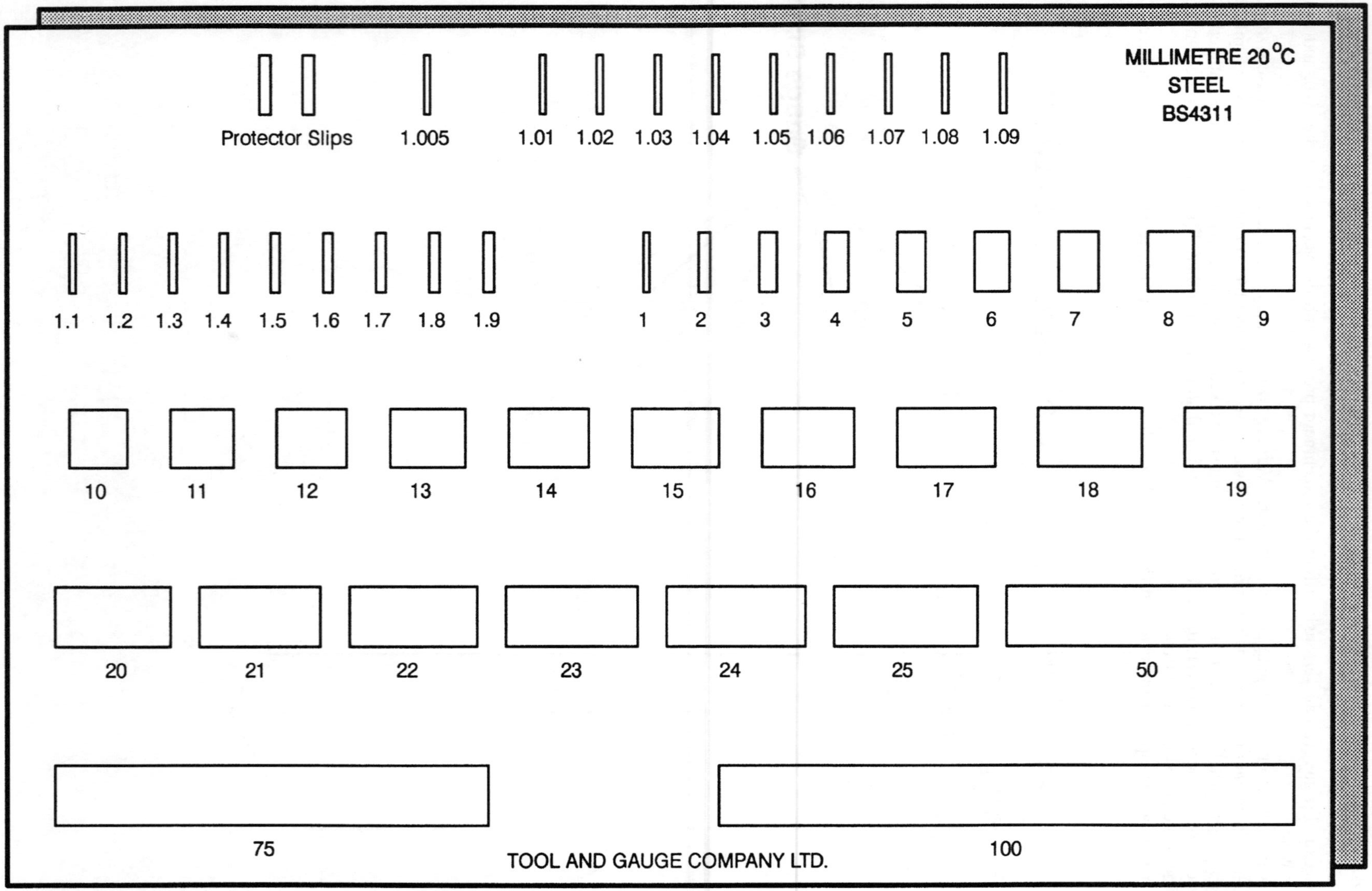
MILLIMETRE 20 °C
STEEL
BS4311
Protector Slips
1.005
1.01
1.02
1.03
1.04
1.05
1.06
1.07
1.08
1.09
1.1
1.2
1.3
1.4
1.5
1.6
1.7
1.8
1.9
1
2
3
4
5
6
7
8
9
10
11
12
13
14
15
16
17
18
19
20
21
22
23
24
25
50
75
100
TOOL AND GAUGE COMPANY LTD.

## 8.5 Sheet metal work

### 8.5.1 Self-secured joints

**Grooved Seam**

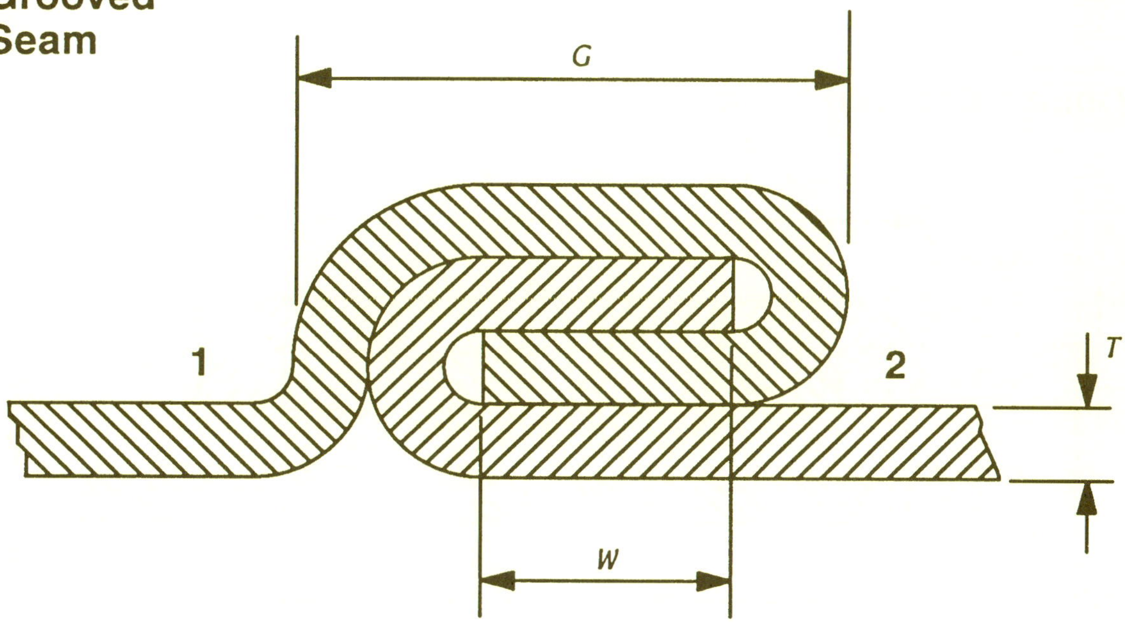


*W* = Width of folded edge (lock)
*G* = Width of grooved seam
*T* = Thickness of metal

Total allowance = $3G-4T$ shared either equally between limbs 1 and 2
or
2/3 of limb 1 and 1/3 of limb 2 where the joint centre position is critical.

## Double Grooved Seam

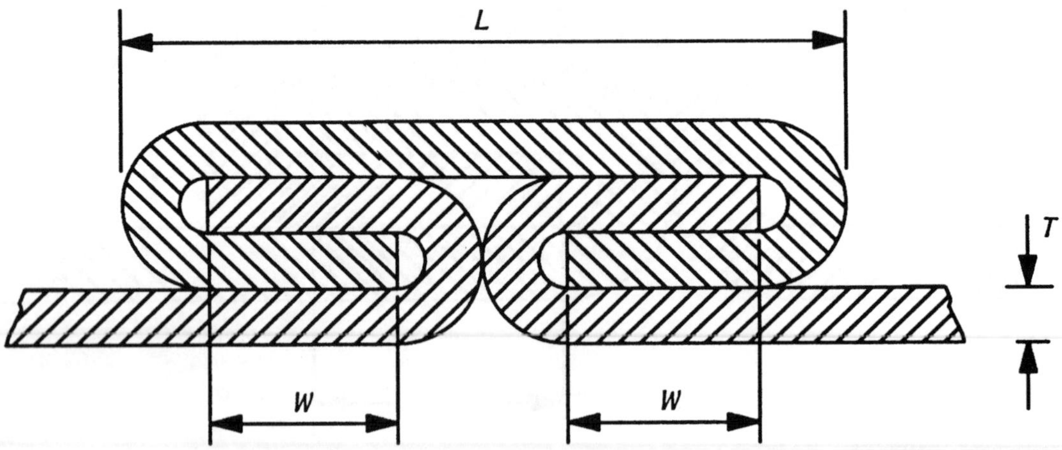


$W$ = Width of folded edge (lock)
$L$ = Width of locking strip
$T$ = Thickness of metal

Add $W-T$ to the edge of each blank to be joined.
The allowance for a capping strip is $4W+4T$ where $L=2W+4T$

# Knocked Up Seam

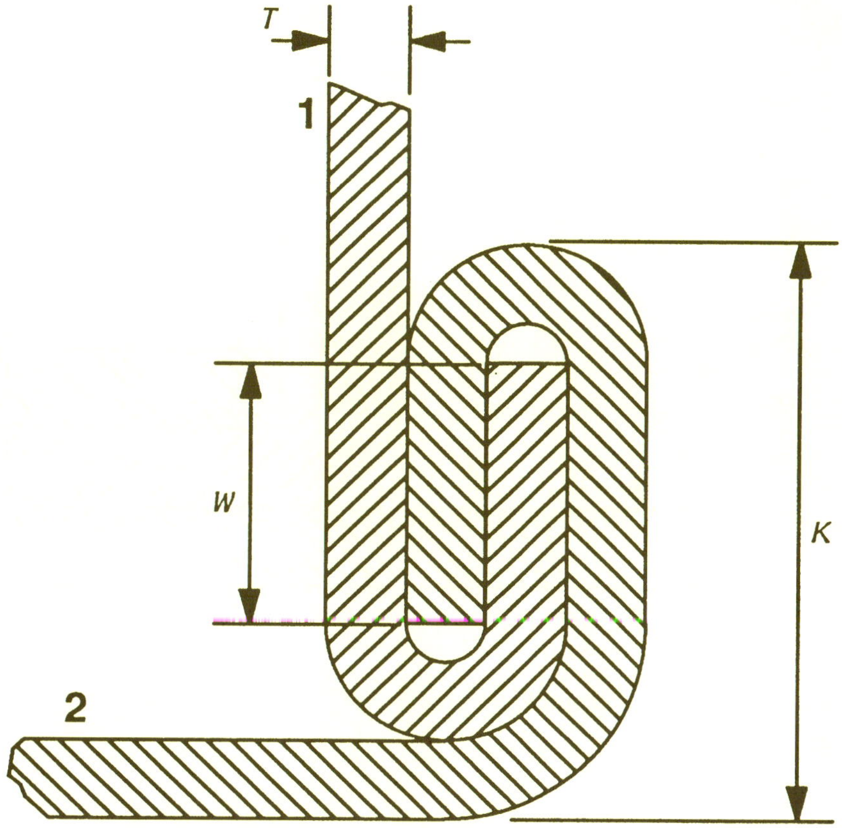


$W$ = Width of folded edge (lock)
$K$ = Width of knocked up seam
$T$ = Thickness of metal

Add $W$ to the single edge 1
Add $2W + T$ to the double edge 2
$K = 2W + 3T$

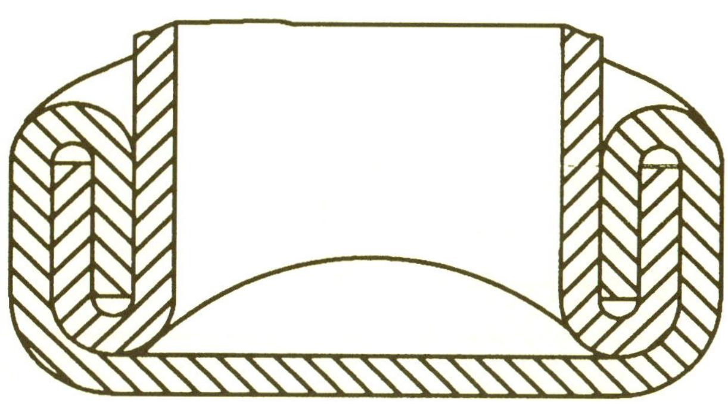

# Paned Down Seam

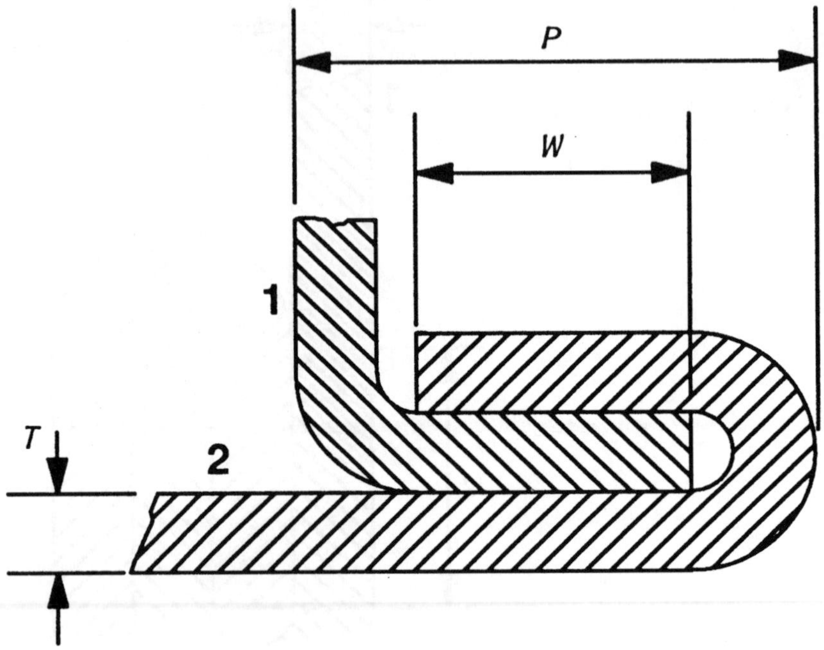


$W$ = Width of folded edge (lock)
$P$ = Width of the paned down seam
$T$ = Thickness of metal

Add $W$ to the single edge 1
Add $2W + T$ to the double edge 2
$P = 2W + 2T$

### 8.5.2 Making a grooved seam

1 Fold edges to form 'locks'.

2 Interlock edges.

3 Close or lock seam using grooving tool to the correct width.

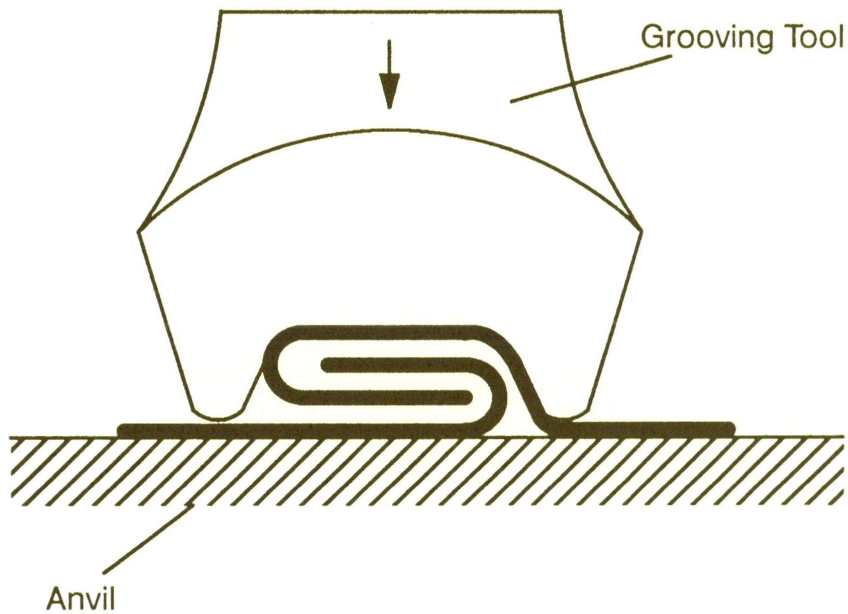

## 8.6 Sheet metal gauge comparisons

| *Gauge Reference Number* | *Standard Wire Gauge – SWG (in)* | *Millimetres Equivalent (mm)* | *Birmingham Wire Gauge (in)* | *American Browne & Sharp Gauge (in)* | *B.G. Sheet and Hoop Iron Standard Gauge (in)* |
|---|---|---|---|---|---|
| 7/0 | 0.500 | 12.669 | – | – | – |
| 6/0 | 0.464 | 11.785 | – | 0.5800 | – |
| 5/0 | 0.432 | 10.972 | – | 0.5165 | – |
| 4/0 | 0.400 | 10.159 | 0.454 | 0.4600 | – |
| 3/0 | 0.372 | 9.448 | 0.425 | 0.40964 | 0.5000 |
| 2/0 | 0.348 | 8.839 | 0.380 | 0.3648 | 0.4452 |
| 0 | 0.324 | 8.229 | 0.340 | 0.32486 | 0.3964 |
| 1 | 0.300 | 7.620 | 0.300 | 0.2893 | 0.3532 |
| 2 | 0.276 | 7.010 | 0.284 | 0.25763 | 0.3147 |
| 3 | 0.252 | 6.400 | 0.259 | 0.22942 | 0.2804 |
| 4 | 0.232 | 5.892 | 0.238 | 0.20431 | 0.2500 |
| 5 | 0.212 | 5.384 | 0.220 | 0.18194 | 0.2225 |
| 6 | 0.192 | 4.876 | 0.203 | 0.16202 | 0.1981 |
| 7 | 0.176 | 4.470 | 0.180 | 0.14428 | 0.1764 |
| 8 | 0.160 | 4.064 | 0.165 | 0.12849 | 0.1570 |
| 9 | 0.144 | 3.657 | 0.148 | 0.11443 | 0.1398 |
| 10 | 0.128 | 3.251 | 0.134 | 0.10189 | 0.1250 |
| 11 | 0.116 | 2.946 | 0.120 | 0.090742 | 0.1113 |
| 12 | 0.104 | 2.641 | 0.109 | 0.0808 | 0.0991 |
| 13 | 0.092 | 2.336 | 0.095 | 0.07196 | 0.0882 |
| 14 | 0.080 | 2.032 | 0.083 | 0.06408 | 0.0785 |
| 15 | 0.072 | 1.828 | 0.072 | 0.05706 | 0.0699 |
| 16 | 0.064 | 1.625 | 0.065 | 0.05082 | 0.0625 |
| 17 | 0.056 | 1.422 | 0.058 | 0.04525 | 0.0556 |
| 18 | 0.048 | 1.218 | 0.049 | 0.0403 | 0.0495 |
| 19 | 0.040 | 1.016 | 0.042 | 0.0358 | 0.044 |
| 20 | 0.036 | 0.914 | 0.035 | 0.0319 | 0.0392 |
| 21 | 0.032 | 0.8124 | 0.032 | 0.0284 | 0.0349 |
| 22 | 0.028 | 0.7109 | 0.030 | 0.0253 | 0.03125 |
| 23 | 0.024 | 0.6093 | 0.025 | 0.0225 | 0.02782 |
| 24 | 0.022 | 0.5585 | 0.022 | 0.0201 | 0.02476 |
| 25 | 0.020 | 0.5078 | 0.020 | 0.0179 | 0.02204 |
| 26 | 0.018 | 0.4570 | 0.018 | 0.0159 | 0.01961 |
| 27 | 0.0164 | 0.4162 | 0.016 | 0.0142 | 0.01745 |

| *Gauge Reference Number* | *Standard Wire Gauge – SWG (in)* | *Millimetres Equivalent (mm)* | *Birmingham Wire Gauge (in)* | *American Browne & Sharp Gauge (in)* | *B.G. Sheet and Hoop Iron Standard Gauge (in)* |
|---|---|---|---|---|---|
| 28 | 0.0148 | 0.3755 | 0.014 | 0.0126 | 0.015625 |
| 29 | 0.0136 | 0.3450 | 0.013 | 0.0112 | 0.013900 |
| 30 | 0.0124 | 0.3146 | 0.012 | 0.0100 | 0.012300 |
| 31 | 0.0116 | 0.2940 | 0.010 | 0.0089 | 0.01100 |
| 32 | 0.0108 | 0.2743 | 0.009 | 0.0079 | 0.00980 |
| 33 | 0.0100 | 0.2539 | 0.008 | 0.0070 | 0.00870 |
| 34 | 0.0092 | 0.2330 | 0.007 | 0.0063 | 0.00770 |
| 35 | 0.0084 | 0.2131 | 0.005 | 0.0056 | 0.00690 |
| 36 | 0.0076 | 0.1937 | 0.004 | 0.0050 | 0.00610 |
| 37 | 0.0068 | 0.1727 | – | 0.0044 | 0.00540 |
| 38 | 0.0060 | 0.1523 | – | 0.0039 | 0.00480 |
| 39 | 0.0052 | 0.1329 | – | 0.0035 | 0.00430 |
| 40 | 0.0048 | 0.1219 | – | 0.0031 | 0.00386 |
| 41 | 0.0044 | 0.1118 | – | – | 0.00343 |
| 42 | 0.0040 | 0.1015 | – | – | 0.00306 |
| 43 | 0.0036 | 0.0914 | – | – | 0.00272 |
| 44 | 0.0032 | 0.0813 | – | – | 0.00242 |
| 45 | 0.0028 | 0.0713 | – | – | 0.00215 |
| 46 | 0.0024 | 0.0610 | – | – | 0.00192 |
| 47 | 0.0020 | 0.0507 | – | – | 0.00170 |
| 48 | 0.0016 | 0.0406 | – | – | 0.00152 |
| 49 | 0.0012 | 0.0305 | – | – | 0.00135 |
| 50 | 0.0010 | 0.0253 | – | – | 0.00120 |
| 51 | – | – | – | – | 0.00107 |
| 52 | – | – | – | – | 0.00095 |

# 9 Materials guide

## 9.1 Properties of materials

### 9.1.1 Properties of solids

Crystal structure (CS): BCC body-centred cubic
FCC face centred cubic
CPH close-packed hexagonal

$\rho$ Mass density (kg/litre) ~ Specific Gravity
$t_m$ Melting point (°C)
$t_b$ Boiling point (°C)
$h_{if}$ Latent heat of fusion (kJ/kg)
$h_{fg}$ Latent heat of vaporization (kJ/kg)
$k$ Thermal conductivity at 0°C (W/mK)
$c$ Specific heat capacity at 0°C (J/kgK)
$\alpha$ Coefficient of linear thermal expansion ($10^{-6}$/K)
$\rho_e$ Electrical resistivity at 20°C (units as shown in tables)

#### 9.1.1.1 Metallic elements

| *Metal* | *CS* | $\rho$ | $t_m$ | $t_b$ | $h_{if}$ | $h_{fg}$ | $k$ | $c$ | $\alpha$ | $\rho_e$ (nΩm) |
|---|---|---|---|---|---|---|---|---|---|---|
| Aluminium | FCC | 2.7 | 660 | 2400 | 387 | 9460 | 205 | 880 | 23 | 27 |
| Copper | FCC | 8.96 | 1083 | 2580 | 205 | 5230 | 390 | 380 | 17 | 16.8 |
| Gold | FCC | 19.3 | 1063 | 2660 | 66 | 1750 | 310 | 145 | 14 | 23 |
| Iron | BCC/FCC | 7.9 | 1535 | 2900 | 270 | 6600 | 76 | 437 | 12 | 97 |
| Lead | FCC | 11.3 | 327 | 1750 | 24 | 850 | 35 | 126 | 29 | 206 |
| Nickel | FCC | 8.9 | 1453 | 2820 | 305 | 5850 | 91 | 444 | 13 | 68 |
| Platinum | FCC | 21.5 | 1769 | 3800 | 113 | 2400 | 69 | 125 | 9 | 106 |
| Silver | FCC | 10.5 | 961 | 2180 | 105 | 2330 | 418 | 232 | 19 | 16 |
| Tantalum | BCC | 16.6 | 3000 | 5300 | 160 | – | 54 | 140 | 6 | 135 |
| Tin | Diamond/tetragonal | 7.3 | 232 | ~2500 | 59 | 2400 | 64 | 224 | 23 | 120 |
| Titanium | CPH/BCC | 4.5 | 1680 | 3300 | 435 | – | 17 | 500 | 9 | 550 |
| Tungsten | BCC | 19.3 | 3380 | ~6000 | 185 | – | 190 | 130 | 4.5 | 55 |
| Zinc | CPH | 7.1 | 420 | 907 | 110 | 1750 | 113 | 384 | 31 | 59 |

### *9.1.1.2 Alloys and non-metals*

| *Type* | *Metal* | $\rho$ | $t_m$ | $k$ | $c$ | $\alpha$ | $\rho_e$ (n$\Omega$m) |
|---|---|---|---|---|---|---|---|
| Alloys | Brass (65/35) | 8.45 | 927 | 120 | 370 | 20 | 69 |
| | Constantan (60/40) | 8.9 | 1320 | 22 | 410 | | 490 |
| | Dural (4.4% Cu) | 2.8 | 640 | 150 | 900 | 23 | ~52 |
| | Manganin (84% Cu) | 8.5 | | 22 | 405 | | 440 |
| | Nichrome (80/20) | 8.36 | | 13 | 430 | 12.5 | 1030 |
| | Phosphor-bronze | 8.92 | ~1050 | ~75 | 380 | 18 | 115 |
| | Mild steel | ~7.85 | | ~50 | ~450 | ~11 | ~120 |
| Non-metals | Alumina | 3.9 | 2050 | 21 | 1050 | 8 | $10^3$–$10^6$ |
| | Brick | 1.4–2.2 | – | 0.4–0.8 | 800 | 3–9 | 1.0–2.0 |
| | Concrete | 2.4 | – | 1.0–1.5 | 1100 | 10–14 | – |
| | Dry ground | ~1.6 | | – | – | – | 0.01–0.1 |
| | Glass | 2.4–3.5 | ~1100 | 0.4–1.1 | 500–800 | 3–10 | $5 \times 10^3$–$10^6$ |
| | Granite | 2.7 | – | 2.0–4.0 | 800 | 6–9 | – |
| | Mica | 2.8 | – | ~0.5 | 840 | – | $10^5$–$10^9$ |
| | Nylon 6 | 1.14 | 200–220 | 0.25–0.33 | 1600 | 80–130 | $10^4$–$10^7$ |
| | Paper (dry) | ~1.0 | – | 0.06 | – | – | $10^4$ |
| | Perspex | 1.2 | 85–115* | 0.19–0.23 | 1450 | 50–80 | – |
| | Polystyrene | 1.06 | 80–105* | 0.08–0.2 | 1300 | 60–80 | $10^{10}$ |
| | Polythene | 0.93 | 65–130* | 0.25–0.5 | 2200 | 110–220 | $10^5$ |
| | PTFE | 2.2 | – | 0.23–0.27 | 1050 | 90–130 | $10^9$ |
| | PVC (Plasticized) | 1.7 | 70–80* | 0.16–0.19 | – | 50–250 | $10^4$–$10^7$ |
| | Porcelain | 2.4 | 1550 | 0.8–1.85 | 1100 | 2.2 | $10^4$–$10^7$ |
| | Quartz (crystal) | 2.65 | – | 5.0–9.0 | 730 | 7.5–13.7 | $10^6$–$2 \times 10^8$ |
| | Quartz (fused) | 2.2 | – | 1.3 | 840 | 0.5 | $10^{10}$ |
| | Rubber | 1.1–1.2 | 125 | ~0.15 | | ~200 | ~$10^7$ |
| | Sandstone | 2.4 | – | 1.1–2.3 | 900 | 5–12 | – |
| | Wood (along grain) | 0.4–0.8 | – | ~0.15 | ~1600 | 3–5 | – |

* Softening temperature

Note that the values given for some of the parameters, in particular $\sigma_y$ and $\sigma_f$ depend on the manufacturing processes used to make the material and its condition. The ranges quoted are typical but not exhaustive. The metals referred to are drawn or wrought unless otherwise stated.

### *9.1.1.3 Mechanical properties of solids*

| *Type* | *Description* | *E* | *G* | *K* | $\vartheta$ | $\sigma_y$ | $\sigma_f$ |
|---|---|---|---|---|---|---|---|
| Metals | Aluminium | 70 | 26 | 75 | 0.34 | 30–140 | 60–160 |
| | Copper | 124 | 46 | 130 | 0.35 | 47–320 | 300–350 |
| | Gold | 80 | 28 | 167 | 0.42 | 0–210 | 110–230 |
| | Iron (cast) | 115 | 45 | – | 0.25 | – | 140–320 |
| | Iron (wrought) | 195 | 76 | – | 0.29 | 160 | 350 |
| | Lead | 16 | 6 | – | 0.44 | – | 15–18 |
| | Nickel | 205 | 79 | 176 | 0.31 | 140–660 | 480–730 |
| | Platinum | 168 | 61 | 240 | 0.38 | 15–180 | 125–200 |
| | Silver | 76 | 28 | 100 | 0.37 | 55–300 | 140–380 |
| | Tantalum | 186 | – | – | – | – | 340–930 |
| | Tin | 47 | 17 | 52 | 0.36 | 9–14 | 15–200 |
| | Titanium | 110 | 41 | 110 | 0.34 | 200–500 | 250–700 |
| | Tungsten | 360 | 140 | – | – | – | 1000–4000 |
| | Zinc | 97 | 36 | 100 | 0.35 | – | 110–200 |
| Alloys | Brass (65/35) | 105 | 38 | 115 | 0.35 | 62–430 | 330–530 |
| | Constantan (60/40) | 163 | 61 | 157 | 0.33 | 200–440 | 400–570 |
| | Dural (4.4%Cu) | 70 | 27 | 70 | 0.33 | 125–450 | 230–500 |
| | Manganin (84% Cu) | 124 | 47 | – | – | – | 465 |
| | Mumetal (77% Ni) | 220 | – | – | – | – | 540–910 |
| | Nichrome (80/20) | 186 | – | – | – | – | 170–900 |
| | Phosphor-bronze | 100 | – | – | 0.38 | 110–670 | 330–750 |
| | Mild steel | 210 | 81 | 170 | 0.30 | 240 | 480 |
| | Steel (high yield) | 210 | 81 | 170 | 0.30 | 450 | 600 |
| Non-metals | Alumina | 200–400 | – | – | 0.24 | 140–200 | 1000–2500 |
| | Brick | 10–50 | – | – | – | – | 69–140 |
| | Concrete | 10–17 | – | – | 0.1–0.21 | – | 27–55 |
| | Glass | 50–80 | – | – | 0.2–0.27 | 30–90 | – |
| | Granite | 40–70 | – | – | – | – | 90–235 |
| | Nylon 6 | 1–2.5 | – | – | – | 70–85 | 50–100 |
| | Perspex | 2.7–3.5 | – | – | – | 50–75 | 80–140 |
| | Polystyrene | 2.5–4.0 | – | – | – | 35–60 | 80–110 |
| | Polythene | 0.1–1.0 | – | – | – | 7–38 | 15–20 |
| | PTFE | 0.4–0.6 | – | – | – | 17–28 | 5–12 |
| | PVC (plasticized) | ~0.3 | – | – | – | 14–40 | 75–100 |
| | Rubber | ~0.001–1.0 | – | – | 0.46–0.49 | 14–40 | – |
| | Sandstone | 14–55 | – | – | – | – | 30–135 |
| | Wood (along grain) | 8–13 | – | – | – | 20–110 | 50–100 |

*Key to above table*
*E* Young's modulus (kN/mm$^2$); *G* Shear modulus (kN/mm$^2$); *K* Bulk modulus (kN/mm$^2$); $\vartheta$ Poisson's ratio; $\sigma_y$ Yield of proof stress (N/mm$^2$); $\sigma_f$ Ultimate stress (N/mm$^2$)

### 9.1.1.4 Hardness number conversion tables

| Rockwell 'B' number | Rockwell 'C' number | Brinell number | Vickers number | Scleroscope number |
|---|---|---|---|---|
| | 70 | 780 | 1150 | 106 |
| | 68 | 745 | 1050 | 100 |
| | 66 | 712 | 960 | 95 |
| | 64 | 682 | 885 | 91 |
| | 62 | 653 | 820 | 87 |
| | 60 | 627 | 765 | 84 |
| | 58 | 601 | 717 | 81 |
| | 57 | 578 | 675 | 78 |
| 120 | 55 | 555 | 633 | 75 |
| 119 | 53 | 534 | 598 | 72 |
| 119 | 52 | 514 | 567 | 70 |
| 117 | 50 | 495 | 540 | 67 |
| 117 | 49 | 477 | 515 | 65 |
| 116 | 47 | 461 | 494 | 63 |
| 115 | 45 | 429 | 454 | 59 |
| 114 | 44 | 415 | 437 | 57 |
| 113 | 42 | 401 | 420 | 55 |
| 112 | 41 | 388 | 404 | 54 |
| 110 | 38 | 363 | 375 | 51 |
| 109 | 36 | 341 | 350 | 48 |
| 108 | 34 | 321 | 327 | 45 |
| 107 | 32 | 302 | 305 | 43 |
| 106 | 31 | 293 | 296 | 42 |
| 105 | 30 | 285 | 287 | 40 |
| 104 | 29 | 277 | 279 | 39 |
| 103 | 26 | 262 | 263 | 37 |
| 102 | 24 | 248 | 248 | 36 |
| 100 | 23 | 241 | 241 | 35 |
| 99 | 22 | 235 | 235 | 34 |
| 98 | 21 | 229 | 229 | 33 |
| 97 | 20 | 223 | 223 | 32 |
| 96 | 18 | 217 | 217 | 31 |
| 95 | 16 | 207 | 207 | 30 |
| 94 | 15 | 202 | 202 | 30 |
| 93 | 13 | 197 | 197 | 29 |
| 92 | 12 | 192 | 192 | 28 |
| 90 | 9 | 183 | 183 | 27 |
| 88 | 7 | 174 | 174 | 26 |
| 86 | 4 | 166 | 166 | 25 |
| 84 | 2 | 159 | 159 | 24 |
| 82 | | 153 | 153 | 23 |
| 80 | | 146 | 146 | 22 |
| 78 | | 140 | 140 | 21 |
| 76 | | 134 | 134 | 21 |
| 74 | | 131 | 131 | 20 |
| 72 | | 126 | 126 | |
| 70 | | 121 | 121 | |
| 68 | | 116 | 116 | |
| 66 | | 112 | 112 | |
| 64 | | 107 | 107 | |
| 62 | | 105 | 105 | |
| 60 | | 101 | 101 | |
| 59 | | 99 | 99 | |
| 57 | | 97 | 97 | |
| 56 | | 95 | 95 | |

*9.1.1.5 Typical compositions of some common alloys*

| *Name* | *Al* | *Cu* | *Au* | *Fe* | *Pb* | *Ni* | *Mg* | *Ag* | *Sn* | *Zn* | *Sb* | *Cr* | *Other* | *Mn* | *P* | *C* | *Si* | *Melting point (°C)* |
|---|---|---|---|---|---|---|---|---|---|---|---|---|---|---|---|---|---|---|
| Aluminium bronze | 10 | 89 | – | 1 | – | – | – | – | – | – | – | – | – | – | – | – | – | 1050 |
| Amalgam (dental) | – | 2 | – | – | – | – | – | 33 | 12.5 | 0.5 | – | – | 52Hg | – | – | – | – | – |
| Babbit | – | 3 | – | – | – | – | – | – | 90 | – | 7 | – | – | – | – | – | – | 241 |
| Battery plate | – | – | – | – | 94 | – | – | – | – | – | 6 | – | – | – | – | – | – | 300 |
| Bearing bronze | – | 82 | – | – | – | – | – | – | 16 | 2 | – | – | – | – | – | – | – | 1000 |
| Constantan | – | 55 | – | – | – | 45 | – | – | – | – | – | – | – | – | – | – | – | 1290 |
| Dural | 94.5 | 4 | – | – | – | – | 0.5 | – | – | – | – | – | – | 0.6 | – | – | 0.4 | – |
| German silver | – | 60 | – | – | – | 15 | – | – | – | 25 | – | – | – | – | – | | – | 1110 |
| Gold (8 carat) | – | 47 | 33 | – | – | – | – | 20 | – | – | – | – | – | – | – | – | – | – |
| Gold (white) | – | – | 82 | – | – | 10 | – | – | – | 8 | – | – | – | – | – | – | – | – |
| Solder (50/50) | – | – | – | – | 50 | – | – | – | 50 | – | – | – | – | – | – | – | – | 225 |
| Leaded red brass | – | 85 | – | – | 5 | – | – | – | 5 | 5 | – | – | – | – | – | – | – | – |
| Manganese bronze | – | 95 | – | – | – | – | – | – | – | – | – | – | – | 5 | – | – | – | 1060 |
| Marine babbit | – | – | – | – | 72 | – | – | – | 21 | – | 7 | – | – | – | – | – | – | – |
| Medal bronze | – | 92 | – | – | – | – | – | – | 6 | 2 | – | – | – | – | – | – | – | 1050 |
| Mechanite | – | – | – | 73–97 | – | – | – | – | – | – | – | – | – | – | – | 2–3 | 1–24 | 1230 |
| Monel metal | – | 33 | – | 6.5 | – | 60 | – | – | – | – | – | 0.5 | – | – | – | – | – | 1360 |
| Mumetal | – | 5 | – | 20 | – | 74 | – | – | – | – | – | – | – | 1 | – | – | – | – |
| Nichrome (casting) | – | – | – | 25 | – | 60 | – | – | – | – | – | 15 | – | – | – | – | – | 1350 |
| Nickel silver (18%A) | – | 65 | – | – | – | 18 | – | – | – | 17 | – | – | – | – | – | – | – | 1110 |
| Phosphor bronze | – | 95.5 | – | – | – | – | – | – | 4.3 | – | – | – | – | – | 0.2 | – | – | 1050 |
| Pewter (clockwork) | – | 61 | – | – | 0.6 | – | – | – | 1.4 | 37 | – | – | – | – | – | – | – | – |
| Red brass | – | 92 | – | – | – | – | – | – | – | 8 | – | – | – | – | – | – | – | 1030 |
| Silicon bronze | – | 97 | – | – | – | – | – | – | – | – | – | – | – | – | – | – | 3 | 1024 |
| Steel (carbon) | – | – | – | 99 | – | – | – | – | – | – | – | – | – | – | – | 1 | – | 1430 |
| Stellite 21 (Co–Cr–W alloy) | – | – | – | – | – | 2 | – | – | – | – | 27 | – | Co + W | – | – | – | – | 1250 |
| Type metal | – | 1 | – | – | 58 | – | – | – | 26 | – | 15 | – | – | – | – | – | – | – |
| White metal | – | 1 | – | – | 75 | – | – | – | 5 | – | 19 | – | – | – | – | – | – | 238 |
| White metal | – | 6 | – | – | – | – | – | – | 82 | – | 12 | – | – | – | – | – | – | 420 |
| Wrought iron | – | – | – | 98.5 | – | – | – | – | – | – | – | – | – | – | – | – | – | 1510 |
| Yellow brass | – | 67 | – | – | – | – | – | – | – | 33 | – | – | – | – | – | – | – | 940 |
| Zamak 5 | 4 | – | – | – | – | – | 0.05 | – | – | 95 | – | – | – | – | – | – | – | 386 |

## 9.2 Cast irons

### 9.2.1 Introduction

Cast irons are ferrous metals containing more than 1.7% carbon, which is the maximum amount which can be held in solid solution in steels.

The amount and form of the carbon present is a key to understanding the cast iron family and descriptive names used appear helpful, e.g. grey, white, ductile, malleable, spheroidal, graphite. Difficulties arise, however, due to the effects of small amounts of other elements such as silicon, sulphur, manganese and phosphorous as well as to the less helpful words which describe the various resulting crystal/grain structures, e.g. cementite, ferrite, pearlite and austenite, many of which are shared with the metallurgy of steels. As with steels, the rate of cooling experienced by a cast iron can have a profound affect on its structure. In other words, two identical castings, cast from the same ladle, may exhibit greatly differing properties if one is cooled through solidification to room temperature substantially faster than the other. Similarly, iron poured into a thin section mould will have different end properties than if it had been cast into a thicker section; the latter will experience slower cooling. Basically the differences which arise are due to the form and distribution of the carbon content. Carbon can either be taken up as combined carbon (cementite) or precipitated out in the form of graphite flakes (free carbon or uncombined carbon); in practice, both forms are usually present.

With slow cooling most, if not all, of the carbon will precipitate out as large graphite flakes; these give the characteristic grey colour to a fractured surface. With faster cooling, some combined carbon will be present and the graphite flakes will be smaller and closer together. As the rate of cooling increases so does the proportion of combined carbon, the free carbon flakes becoming smaller and closer. With very rapid cooling, no free carbon is apparent on a fractured surface: hence the term 'white' cast iron.

Silicon present in a cast iron tends to increase the proportion of free carbon while sulphur has the opposite tendency. Manganese combines with sulphur and floats to the surface of the melt to join the slag but it also makes combined carbon more stable. Phosphorous has little direct effect on the proportion of combined to free carbon; it improves fluidity but also forms compounds which are hard and brittle.

### 9.2.2 Grey cast iron (flake graphite)

Seven grades of *grey cast iron* are specified with tensile strengths stepping from 150 N/mm$^2$ to 400 N/mm$^2$. Compressive strengths are substantially higher since they are little affected by the discontinuities represented by the graphite flakes. It is these flakes which cause the stress/strain curve to be non-linear, giving values of modulus of plasticity, elasticity and rigidity from one-third to two-thirds of those of steel as the proportion of free graphite decreases.

### 9.2.3 White cast iron

There is no free graphite in *white cast irons*, so these are very hard and brittle with few direct applications. The white structure is promoted by rapid cooling or local chilling in grey cast iron: this provides a means of local surface hardening but must be avoided in other circumstances. White cast iron is the starting stage for malleable iron.

### 9.2.4 Ductile cast irons

*Ductile cast irons* comprise malleable and nodular (or spheroidal) graphite irons.

#### 9.2.4.1 Malleable irons (whiteheart, blackheart and pearlite)

*Malleable irons* are produced by heat-treating white cast irons to form free graphite as aggregates or nodules dispersed throughout the metallic matrix. They have good tensile strengths, with high proof stress to tensile strength ratios as well as substantial ductility and shock resistance. Under stress they behave in a manner similar to steel with a straight stress/strain curve, indicating a constant modulus of elasticity.

*Whiteheart malleable cast iron* is made by heat-treating a white cast iron under decarbonizing conditions. Thin sections may be almost completely decarbonized and may have properties similar to those of mild steel.

*Blackheart malleable cast iron* is made by heat-treating white cast iron, avoiding decarbonization. The whole combined carbon content is converted into graphite in the form of temper carbon aggregates. This type of cast iron also has good ductility and shock resistance coupled with very good machinability.

*Pearlitic malleable cast iron* is produced in a similar way to the blackhearted version but a pearlitic matrix is achieved either by the addition of carbide stabilizing elements to the original white cast iron or by a two-stage heat treatment. The resulting cast iron is characterized by higher tensile strength than other malleable irons, together with lower elongations and higher hardness. It machines readily and can easily be induction- or surface-hardened.

#### 9.2.4.2 Nodular or spheroidal graphite cast irons

Uncombined graphite appears throughout the as-cast structure as spherical particles as a result of treatment of the metal by the addition of magnesium or cerium before pouring. The steel-like matrix can be modified by annealing for high impact strength and ductility or by normalizing for higher tensile strength with lower ductility. *Nodular cast irons* have high tensile strengths, a high elastic modulus and considerable ductility together with corrosion resistance and castability comparable to that of flake graphite cast iron. The mechanical properties are much less sensitive to section thickness (cooling rate) than grey cast iron.

Table 9.2.1 gives an outline of the key mechanical properties of the major ductile cast irons.

**Table 9.2.1** *Typical mechanical properties of key cast irons*

| *Type* | *Tensile strength ($N/mm^2$)* | *Elongation (%)* | *Elastic modulus ($KN/mm^2$)* | *Hardness (HB)* |
|---|---|---|---|---|
| Whiteheart | 340–410 | 3–4 | 176 | 120–180 |
| Blackheart | 290–340 | 6–12 | 169 | 130–150 |
| Pearlitic | 440–690 | 7–2 | 172 | 149–285 |
| Nodular | 370–800 | 17–2 | 169–176 | 116–350 |

### 9.2.5 Special purpose cast irons

Other cast irons have been developed for particular service conditions, such as corrosion resistance, low thermal expansion, non-magnetic properties, high temperature resistance, high strength with high wear resistance, high hardness with high abrasion resistance, resistance to corrosion, creep, growth and scaling at high temperatures, and so on. Most of these requirements are specific to particular industries, such as the petro-chemical industry, therefore particular irons find particular niches.

## 9.3 Non-ferrous metals and alloys

### 9.3.1 Introduction to terminology and definitions

*Non-ferrous metals* are the metallic elements, excluding iron. Many of these elements are alloyed with iron to form various types of steel and cast irons: these ferrous alloys are not considered in this section.

Few metals are used in engineering in their (near) pure form and even then only for particular physical properties, such as electrical and thermal conductivity (e.g. copper and aluminium) and corrosion resistance (e.g. silver and gold for electrical contacts; zinc, tin and cadmium for corrosion protection). Alloying can produce greatly enhanced mechanical, processing and other properties. For example, 0.1% silver added to copper raises its annealing temperature by 150°C, allowing work-hardened components to be soldered without being softened.

Alloys are mixtures of two or more metals which are compatible with each other in the sense that their atoms intersperse and relate to each other in an orderly manner. This contrasts with mixtures of metals which exist within each other without atomic bonding. For example, when lead is added to brass (itself an alloy of copper and zinc) to improve its machinability (Section 9.5.4.14), it exists within the metallic structure of the alloy as isolated globules.

Few non-ferrous alloys can be hardened by heat-treatment (Section 9.5.4.2) but many become strained or work-hardened (Section 9.5.4.9) during processing: this may or may not be an advantage, depending upon the next stage of manufacture. For example cold-rolled (Section 9.5.4.12) aluminium may be 'half hard' or 'hard' depending on the amount of cold work involved: aluminium alloys in a half hard condition machine readily but are not suitable for bending or other forming processes, having much reduced ductility. The condition of an alloy may therefore often be as important as its precise chemical composition. Further, the work hardening characteristics of an alloy may be such as to require inter-stage annealing (Section 9.5.4.10) during manufacturing processes which involve straining the material.

## 9.3.2 Commonly used alloy families

With so many metals available, the permutations, combinations and compositions possible are infinite and indeed there are a vast number of alloys available. In general engineering, however, most required properties for most applications can be met by materials chosen from a relatively few alloy systems. These are now outlined.

### 9.3.2.1 Aluminium alloys

Pure aluminium is light, soft, ductile and corrosion-resistant with high electrical and thermal conductivity. Alloying may improve its strength, hardness and machinability or allow enhanced properties by heat treatment as desired.

Alloying elements include copper, magnesium, silicon, manganese, iron, nickel and zinc, with chromium and titanium also used in small amounts.

Most aluminium alloys have a strong, stable oxide film which reforms virtually instantaneously when broken and which is resistant to many corrosive conditions. Aluminium/magnesium alloys are used in marine applications, as are cast aluminium/silicon alloys. During welding the formation of the oxide film is inhibited by the use of an inert gas shield, e.g. argon in TIG welding.

Aluminium's natural corrosion resistance may be enhanced by processes such as anodizing, a electrolytic treatment of the surface which may include dyeing for aesthetic effects.

*Wrought alloys* are those used in the production of sheet, bars, rolled sections and forgings as opposed to aluminium casting alloys. Commonly used materials are alloys with manganese (1.25%) or magnesium (2.25, 3.5 or 4.5%). These cannot be heat-treated but may be strengthened by cold-working (rolling or drawing) at the cost of ductility. Other alloys, comprising usually more than one other element, can be heat-treated and some can age-harden.

*Cast alloys* have various formulations, not only to provide particular characteristics in the finished component, but also to facilitate the casting process – whether sand casting or die casting. Most contain between 5% and 12% silicon (with a maximum of 23%) with or without other elements such as magnesium, copper, zinc and manganese.

Properties of three commonly used casting alloys are given in Table 9.3.1: actual values depend on casting methods, e.g. sand cast, pressure die cast.

**Table 9.3.1** *Properties of common cast aluminium alloys*

| *Alloy type* | *Alloying elements* | *Tensile strength ($N/mm^2$)* | *Elongation (%)* |
|---|---|---|---|
| LM2 | 88%Al, 10%Si, 2%Cu | 185–247 | 1.5–2 |
| LM4 | 91.5%Al, 5%Si, 3%Cu, 0.5%Mn | 147–170 | 2–3 |
| LM6 | 88.5%Al, 11.5% Si | 170–216 | 4–7 |

*Typical general physical and mechanical properties of aluminium alloys*

Young's modulus 69 $kN/mm^2$
Torsion modulus 26 $kN/mm^2$
Poisson's ratio 0.34
Melting points between 450°C and 660°C
Coefficient of linear expansion $24 \times 10^{-6}/°C$
Tensile strength 50–450 $N/mm^2$ } depending on alloy and condition
Elongation 3–30%

### 9.3.2.2 Copper and copper alloys

*Copper alloys* are a very versatile group of materials characterized by strength, conductivity, corrosion resistance, machinability and ductility, each to varying extents depending on their composition and treatment.

For electrical conductivity applications, small amounts of alloying elements improve specific properties. For example, 0.1% silver allows work-hardened copper to be soldered without softening; cadmium adds strength and toughness for overhead conductors and wires of the type used in railways and telephones; 0.5% tellurium greatly increases machinability. Copper is a relatively expensive metal and is used mostly for electrical applications. For other applications it is alloyed with somewhat less costly metals: thus the less copper in the alloy, the cheaper it tends to be.

*Brasses* are alloys with zinc (between 5% and 40%), sometimes with additional elements to improve properties such as strength, corrosion resistance and machinability. For cold-working processes such as bending and deep drawing, the zinc content may be restricted to 30% for maximum ductility; cartridge and shell cases are common applications.

*Bronzes and gunmetals* are copper–tin alloys with small amounts of other metals such as zinc, phosphorus, nickel and lead; those containing zinc are known as *gunmetals* while those with deliberately controlled amounts of phosphorus are known as phosphor bronzes. *Wrought phosphor bronzes*, with 3–8.5% tin and 0.02–0.34% phosphorus, are used for helical and flat springs, their springiness being induced by work hardening involved in their manufacture. *Cast phosphor bronzes*, with typically 10% tin and 0.5% phosphorus, are used for plain bearing applications.

*Admiralty gunmetal*, containing 88% copper, 10% tin and 2% zinc, is used for valves and steam pipe fittings where strong corrosion-resistant castings are required. Other gunmetals with nickel contents are used where higher strengths and hardness are required to resist wear, erosion and corrosion, typically in valves and pumps.

*Aluminium bronzes*, with 6–11% aluminium and usually lesser amounts of iron, manganese and nickel, are highly resistant to oxidation and are used at higher than ambient temperatures in heat exchangers and condensers in chemical and steam raising plant. Other applications include valve fittings, electrical brush holders, worm gear wheels and heavy duty bearings.

*Typical mechanical properties*

*Note:* All of the values given in Table 9.3.2 depend greatly on the the formulation of the alloy and on its condition, i.e. the degree of strain hardening.

**Table 9.3.2** *Properties of common copper alloys*

| *Copper alloy type* | *Tensile strength ($N/mm^2$)* | *Elongation (%)* | *Hardness (VPN)* |
|---|---|---|---|
| Brasses | 280–730 | 4–70 | 60–185 |
| Bronzes, phosphor bronzes and gunmetals | 140–740 | 2–65 | 30–210 |
| Aluminium bronzes | 385–775 | 3–70 | 80–250 |

### 9.3.2.3 Magnesium alloys

Although *magnesium* is a relatively weak metal and burns in air at temperatures not much in excess of its melting point, its alloys have useful lightweight properties (i.e. good strength to weight ratio since its relative density is 1.8 compared with 2.7 for aluminium and 7.9 for steel). Most applications are confined to the aerospace industries: typical examples are landing wheels, crank-cases and engine parts.

Typical cast alloys contain 8–10% Al, 0.3% Mn, 0.7–4.5% Zn and 0.7% Zr with tensile strengths 139–263 $N/mm^2$ and associated elongation 2–7% (zirconium is an effective grain refiner).

Wrought alloys contain 3–6% Al, 0.3–1.5% Mn, 1–3% Zn and 0.7% Zr with tensile strengths 201–309 $N/mm^2$ and associated elongation 5–8%.

### 9.3.2.4 Zinc alloys

*Zinc* is widely used in engineering for corrosion protection of ferrous alloys. *Zinc alloys*, on the other hand, are used almost exclusively for high pressure die cast components because of their excellent casting characteristics and the relatively high strength imparted to components by the chilling effect of the metallic mould (or die). Since dies are expensive to make they must be paid back over a long production run so zinc components are confined to the mass production end of the manufacturing spectrum.

Only two or three alloys are used and differ from each other in the amount of copper included in the alloy – this varies from 0 to 2.7% – otherwise the balance is 4.1% aluminium and the rest zinc. Other metals are regarded as contaminates because even very small amounts can have devastating effects on the properties of the alloy: great care is therefore taken to minimize or eradicate magnesium, iron, lead, cadmium and tin.

*Typical physical and mechanical properties*

| | |
|---|---|
| Tensile strength | 283–332 N/mm$^2$ |
| Compressive strength | 409–593 N/mm$^2$ |
| Elongation | 9–15% |
| Hardness | 83–92 VPN |
| Thermal expansion | $27 \times 10^{-6}$ mm/°C |
| Melting range | 379–388°C |
| Specific gravity | 6.7 |

#### 9.3.2.5 Nickel chromium alloys

*Nickel chromium alloys* are used where their ability to resist oxidation at high temperatures is a desired characteristic. In addition, those with high electrical resistance are used for heater elements. Perhaps the most exacting applications are in the jet and rocket engine fields but more mundane uses are found in furnaces, thermocouples, food processing equipment, etc. There are also alloys designed for more ordinary temperatures in the chemical industries where resistance to attack by acids is required.

Some nickel alloys drop the partnership with chromium in favour of molybdenum when hydrochloric, phosphoric and other non-oxidizing acids are involved.

## 9.4 Plastics

### *9.4.1 Introduction*

*Plastics* is the name popularly given to polymeric materials, although many of them only have plastic properties when heated to their processing temperatures for moulding, extrusion, etc.; as finished components they often exhibit elastic properties.

*Polymers* take their name from the long chain (polymer) molecules from which they are formed. Most fall into either the *thermo-plastic* or *thermo-setting* plastic categories.

Simple *linear polymer chain* molecules are typical of non-rigid, flexible, thermo-plastic materials such as polythene (polyethylene to give it its proper name). *Branch linear polymer chain* molecules result in thermo-plastic materials which are more rigid, harder and stronger, although less dense; while in *cross-linked polymer* materials the long polymer chains share branches and so are rigid and hard and tend to be brittle – these are the thermo-setting plastics (thermo-sets). *Elastomers*, with their rubber-like properties, have limited cross-linking. Unlike thermo-plastics, which can be returned to a soft, plastic condition by heating, the curing process involved in establishing the cross-links in the thermo-sets is irreversible; subsequent heating above a particular level results only in degradation and charring.

Polymers are amorphous; their long chain molecules are randomly intertwined like worms in a can. Some degree of order, crystallinity, can be introduced within the material; these zones are known as crystallites. Within them, the molecules are folded over each other like paper towels in a dispenser. The degree of a crystallinity is established during processing and can vary from virtually nothing to over 90%. With increased crystallinity comes greater density and stiffness. For example, low density polyethylene (LDPE) is about 60% crystalline compared with the 95% level for the high density version of the same material (HDPE). LDPE is not so strong or tough but is more flexible and ductile. Generally amorphous plastics tend to be transparent and become translucent and then opaque as crystallite regions increase in size and number.

Many polymers are characterized by a glass-transition temperature below which they are hard, brittle and glassy. For some this temperature is above ambient (e.g. Nylon66 at 57°C; polycarbonate at 150°C) so they are generally known in their glassy state, while for others their glass transition temperature is below 0°C (e.g. polypropylene at –14°C; rubber at –73°C). Thus the temperature range over which a component is required to operate obviously has to be taken into consideration: a component made of Nylon66 may operate entirely satisfactorily at room temperature but fail above 57°C, far below its melting point of 265°C.

### 9.4.2 Composite plastic materials

*Glass reinforced plastics* (GRP) are probably the best known of these materials, being encountered in boating, motoring and other leisure and industrial applications. The reinforcing glass fibres may be in the form of randomly orientated chopped strands or, alternatively, woven into a cloth prior to impregnation by the epoxy resin. Other fibres used for reinforcement include aramid ('Kevlar') and carbon but these tend to be expensive – as sportsmen in a number of different fields know all too well.

*Laminated plastics*, where the reinforcement may be cellulose (paper), textile cloth or glass fibre cloth, are also widely used in industry. These materials are usually produced in sheet or bar form and are subsequently machined to produce products such as gears, screws and plain bearings.

### 9.4.3 Fillers and other additives

*Fillers* are added to plastics both to reduce the cost of the finished product and to improve manufacturing and performance properties. For example, fillers such as glass fibre, shredded paper and mica granules reduce shrinkage during moulding as well as improve electrical insulation properties.

Other *additives* include those which improve resistance to degradation by sunlight (ultraviolet radiation) and those which reduce the accumulation of static electricity (with its associated attraction for dust particles).

*Pigments* (organic or inorganic) and *dyes* are used to colour translucent and transparent plastics (respectively).

### 9.4.4 Engineering thermo-plastics

This categorization has arisen in the attempt to distinguish those thermo-plastics which are suitable for long-term loading from those suitable for usually short-term low stress applications, for example in packaging. Although there are many grey areas, materials which are generally recognized to be in this bracket include *polymide*, *acetal*, *polycarbonate*, *polyester*, *polysulphone* and *polyphenylene sulphide*. Other new high temperature engineering thermo-plastics are being developed as are thermo-plastic 'alloys' – combinations of two or three plastics, designed to provide properties not available in a single plastic.

### 9.4.5 The mechanical properties of plastics

Plastic material properties differ in several respects from those of metals. More reliable and effective designs can be achieved if the differences are recognized and understood but often 'effective' values for properties such as Young's modulus and Poisson's ratio are used and plastics then treated as if they were metals.

Unlike metals, the relationship between stress and strain for plastics is dependent on the rate of loading, the frequency of loading and the length of time for which the load is applied (i.e. they exhibit visco-elastic properties, behaving both like viscous liquids and elastic solids). A plastic material may appear to be much stiffer when loaded rapidly than if the load is applied more gradually. However, if the material remains loaded for some time, strain will increase: the material will *creep*. The rate of creep is usually dependent on the ambient temperature. When the load is removed, recovery to the material's original dimensions is also time/temperature dependent. Thus to use the normal strength of materials equations applicable to metals, appropriate values of time-dependent properties have to be determined from the plastic's creep properties.

### 9.4.6 Other properties of plastics

#### 9.4.6.1 Density, strength and stiffness

Absolute values of strength and stiffness of even the 'best' plastics compare unfavourably with those of metals and alloys: it is only when density is taken into account and comparisons of strength and stiffness are made on a basis of weight that the advantages emerge. These become particularly evident when good engineering design is coupled with manufacturing methods which allow the production of shapes of high rigidity using thin sections. Strength and stiffness can be further increased by the use of reinforcement (Section 9.4.2).

### 9.4.6.2 Toughness

The impact resistance of plastics varies from those which are virtually unbreakable to those which shatter like glass. A complicating factor is temperature; some materials which are reasonably tough at room temperature become brittle as temperatures fall below 0°C.

### 9.4.6.3 Fatigue

Conventionally, fatigue arises in engineering components which are subject to variations in stress over many thousands of cycles. This may be more serious in some applications with plastics due to their inherent high internal damping properties coupled with low thermal conductivity which can result in the increase of internal temperature, leading to heat softening failures. Designers aim to reduce the amount of cyclic deformation and improve heat transfer.

The creep to which loaded plastic components are liable (Section 9.4.5) can be regarded as an unconventional 'static' form of fatigue.

### 9.4.6.4 Temperature effects

Although some plastics can function satisfactorily up to 250°C, most have a much lower upper limit. Further, some become brittle below 0°C. Thermo-plastics tend to soften over a range of elevated temperatures whereas thermo-sets will usually still function until they reach temperatures which cause them to char.

### 9.4.6.5 Flammability

Few, if any, thermo-plastics can survive flames and, even if combustion retardants are added, the emission of smoke and toxic fumes can cause problems. Very careful consideration must therefore be given to situations where a fire hazard might exist.

### 9.4.6.6 Chemical inertness

Most plastics, particularly thermo-sets, have excellent resistance to attack by most chemicals but there are exceptions. Data banks of information for all plastic/chemical components now exist and information is available from manufacturers.

### 9.4.6.7 Weather resistance

Exposure to the outdoor environment brings plastics into contact with a combination of factors which may together prove detrimental. Factors include sunlight (ultra-violet (UV) radiation) and atmospheric gases and pollutants, perhaps dissolved in rain water. Some plastics, such as acrylic and PVC (Section 9.4.6.10), perform well, while others, such as polystyrene and cellulose acetate, suffer rapid degradation.

### 9.4.6.8 Coefficient of friction

PTFE (Section 9.4.6.10) is renowned for its slipperiness, but most plastics have a coefficient of friction in the range 0.3–0.5 when sliding on metal. For bearing applications, PTFE and other low-friction plastics lack stiffness and are therefore often used to coat or impregnate other materials, such as porous bronze. This type of application has the additional advantage of improved thermal conductivity and hence greatly reduced risk of the low friction properties of the bearing being destroyed by overheating. The coefficient of friction of other plastics can be reduced by the addition of fillers, e.g. graphite in phenol and molybdenum disulphide in nylon.

### 9.4.6.9 Electrical resistance and dielectric strength

Plastics are good insulators, although some may absorb moisture from the atmosphere in humid conditions with a consequent reduction in resistance. Similarly, they have high dielectric strengths. These properties can be detrimental in that charges of static electricity can accumulate, attracting dust and also giving rise to sparks (which must be avoided in an explosive or inflammable atmosphere). Anti-static coatings or additives are sometimes necessary.

### 9.4.6.10 Thermo-plastics

In this section some of the most common thermo-plastics are described along with their applications.

**Acrylonitrile – Butadiene – Styrene (ABS)** This is a very tough, hard and rigid plastic with good abrasion resistance and low water absorption. It has good dimensional stability and can be electro-plated. It is susceptible to attack by a variety of chemicals. Typical uses include computer housings, fascias, helmets and luggage.

**Acetal** This plastic has a low coefficient of friction, very good dimensional stability and resistance to vibration fatigue as it is a very strong and stiff plastic. It has a high resistance to abrasion and chemicals. Used for plumbing, kettles, gears, bearings and conveyor belts, it has a low water absorption and retains most of its properties when immersed in or containing hot water.

**Acrylic** This plastic has a hard glossy surface with high optical clarity, as well as being supplied in a range of transparent colours. It has very good electrical properties and moderate chemical resistance. Polymethyl methacrylate (Perspex) is the most widely known of this family and is used for machine guards, signs and vandal-proof glazing.

**Cellulosics** These are a family of hard and tough material, namely butyrate, cellulose acetate, ethyl cellulose and propionate. A number of compounds are available so it is difficult to give a general outline of available properties. Dimensional stability is usually poor to reasonable and various plastics are available in bright colours.

**Fluoroplastics** Typical plastics are FEP, PFA, PTFEM, ETFE and PVF. They are characterized by having very excellent stability at high temperatures with very good electrical properties and chemical resistance. Although very expensive they are not very strong; however, this is not a characteristic which they would be chosen for.

**Polyamide (Nylon)** Nylons have excellent toughness and wear resistance combined with low friction, strength and stiffness; electrical properties and chemical resistance are very good. Dimensional stability is generally not very good; however, this does depend on the nylon type. The main uses are for bearings, gears, ratchets, rollers, wheels, terminal blocks, etc.

**Polycarbonate** This transparent plastic has excellent resistance to load-induced creep as well as good outdoor stability. It has a very high impact resistance and reasonable chemical resistance. It can be affected by stress cracking due to contact with aromatic solvents. It is used for vandal-proof applications, machine housings and guards, hospital equipment, etc.

**Polyester** This is a tough plastic with excellent dimensional stability and electrical properties. It has outstanding chemical resistance except when exposed to strong acids or bases. Usually it is not used outdoors or for continuous service in water over 52°C. It is commonly used for automobile parts, gears, housings, pulleys, impellers and electrical component parts.

**Polyethylene (PE)** A number of types of plastic exist in this category; they can be low-density (LD), medium-density (MD) and high-density (HD). LD PEs are are usually flexible and tough while MD and HD are stronger, harder and more rigid. Dimensional stability and heat resistance are poor; however, they have outstanding electrical properties and chemical resistance as well as being very lightweight. Applications for these low cost, easily processed materials include detergent and washing liquid bottles, tubing, bins, crates and pipes.

**Polymide** This expensive plastic has very good resistance to heat up to a temperature of 260°C if in use continuously and 483°C intermittently. It has high impact strength as well as good resistance to heat aging and wear. It has a low coefficient of thermal expansion but tends to be quite difficult to process. Usually chosen for high temperature applications as bearing, brushes, insulation, etc.

**Polyphenylene sulphide** Usually reinforced with glass fibre (up to 40%) to counteract inherent brittleness, it has outstanding chemical and heat resistance (up to 230°C) in constant use. It has excellent low-temperature strength as well as high stiffness and creep resistance. It requires a high processing temperature and is inherently flame-retarding. It is resistant to most chemicals over a wide range of temperatures. It is widely used in electrical components, including relays, motor housings and coil formers. Automobile applications include carburettor parts and fluid control valves.

**Polypropylene (PP)** PP has excellent electrical properties and chemical resistance. It has outstanding resistance to flex and stress cracking and above –10°C exhibits good impact strength as well as good thermal stability up to 107°C. It is very lightweight but cheap and is available in a range of grades and as a co-polymer (ethylene/propylene). Usage is widespread and includes small machine and car parts, computer cabinets, tool handles and components with integral hinges. It can be electro-plated.

**Polystyrene (PS)** This is a low cost plastic which is transparent and rigid as well as being easy to work. Grades range from brittle to tough. It tends to have low heat resistance and requires modification to improve heat and impact resistance. It is widely used for food and dairy product containers as well as electrical coils, relays and fittings, but has poor outdoor stability.

**Polysulphone (PSU)** PSUs require high temperature processing and therefore have the highest operational temperature of any thermo-plastic. They are tough, strong and very stiff with excellent electrical properties even at high temperatures, and can be electro-plated. Applications include microwave cooking utensils, fan heaters, medical equipment, hair-dryer parts and printed circuit boards.

**Polytetrafluoroethylene (PTFE)** PTFE is one of the main fluoroplastics (see earlier); it has the lowest coefficient of friction of any solid, it is chemically inert, weather resistant, and has a very high melting temperature (328°C). Expensive and prone to creep, it is highly viscous when molten, making it impossible to process by conventional methods. Powder compacting and sintering is the usual route to bar or sheet material; it is subsequently machined to shape. As well as non-stick coatings, PTFE is used for medical applications such as catheters and in engineering for bushes, bearings and piston rings.

**Polyvinyl chloride (PVC)** This type of plastic is resistant to weather and has good electrical insulation properties. It is also self extinguishing. Plasticized PVC, being flexible, is used for electrical wiring insulation, children's toys and waterproof clothing. Unplasticized PVC, which is hard, tough and strong, is used in buildings for pipes, window frames and guttering.

### *9.4.6.11 Thermo-sets*

**Aminos** (urea formaldehyde and melamine formaldehyde) are hard, rigid and resistant to abrasion. They can also operate at temperatures up to 100°C.

Usage of *urea formaldehyde* is restricted to applications where moisture absorption (and consequent dimensional changes) can be tolerated; examples include electrical switches, plugs and fittings, utensil handles and trays.

*Melamine formaldehyde* has lower moisture absorption and distortion and better resistance to chemicals. It is used for tableware, worktops and electrical fittings.

**Phenolics (phenol-formaldehyde)** has a long history as 'Bakelite'. It is strong, hard, brittle, creep resistant and a good electrical insulator. It has poor resistance to attack by alkalis and oxidizing agents. Applications are widespread in domestic kitchenware; also in fans and pump parts.

**Polyurethanes** are available as rigid foam, flexible foam and elastomer. Strong and resistant to abrasion and chemicals, foams are used for insulation and cushioning and elastomers are used for solid tyres and shock absorbers.

**Polyesters** are widely known to DIY enthusiasts in the form of 'fibreglass kits' and are normally used industrially for tanks and containers.

**Epoxides** are reinforced epoxy resins and are used in the aircraft industry where toughness, resistance to weathering, low moisture absorption and ability to function over a temperature range of –26°C to 150°C are desirable properties.

### *9.4.6.12 Plastic processing*

The main methods of producing plastic components fall into the categories of moulding, extrusion, calendering and forming. Moulded components require little or no subsequent machining while the other methods sometimes only produce bar or sheet material for machining to shape or size.

**Moulding** processes such as blow moulding and rotational moulding produce hollow components by using air pressure or centrifugal force to cover the inside of the mould which has been loaded with a precise amount of plastic material.

**Injection**, compression and transfer moulding are analogous to the casting of metals: the mould cavity is completely filled, internal or hollow features being created by core pins or similar.

**Extrusion** processes convert powdered or granulated feed stock into lengths of material with particular cross-sections (which may be solid or hollow). The extrudate may have been designed to require little more than chopping to length to produce the final article (e.g. curtain rail or window framing) or it may be the starting point for subsequent machining.

**Calendering** rolls molten thermo-plastic into continuous sheets of uniform thickness. It is also used for applying a continuous uniform coating to paper, textiles, etc., mainly for elastomers and PVC.

**Forming** is carried out on sheet plastic material above its glass transition temperature (which, for some plastics, is below normal ambient temperature). The sheet is made to conform to the shape of a mould by hand shaping, or atmospheric pressure (vacuum forming), or compressed air, or by being confined between male and female matched moulds.

## 9.5 Steels

### *9.5.1 Introduction*

There are standard specifications for several hundred steels for a wide range of uses but the majority of engineering applications can be satisfied from a relatively narrow range. Some of these are listed in the tables given in this section.

Fundamentally, all steels contain iron alloyed with carbon, i.e. carbon atoms integrated into the molecular structure of iron. Plain carbon steels are a range of materials with a varying percentage of carbon; this in turn gives a variety of properties. When further alloying elements are added to give enhanced properties the materials are generally known as alloy steels.

### *9.5.2 British Standard specification for wrought steels – BS970*

The revision of BS970 in 1970/72 rationalized the 'En' designations of the 1955 version replacing the En numbers with a five-figure/one-letter code, e.g. En1A became 230M07 and En8C became 080A40. Some En specifications were abandoned and were given no equivalent '1970' designation and some new standard steels were specified, i.e. those which were not directly equivalent to any particular En series. En numbers have lingered on in many industries and are still recognized by steel stockists.

The first three digits identify the family of steels to which the alloy belongs:

(a) 000–199: carbon and carbon manganese steels where the figures indicate 100 times the mean manganese content
(b) 200–240: free-cutting steels where the second and third figures indicate 100 times the sulphur content
(c) 250: silica-manganese spring steels
(d) 300–499: stainless, heat-resisting and valve steels
(e) 500–999: alloy steels in groups or multiples of groups according to alloy type.

Under the current system, the fourth character specifies the steel's condition: 'A' denotes that the steel will be supplied close to the limits of chemical composition; 'H' relates to hardenability; 'M' to specified mechanical properties; 'S' denotes a stainless alloy.

The fifth and sixth digits represent 100 times the mean carbon content of the steel.

Some of the most common steels with their standards' references are given in Table 9.5.1.

**Table 9.5.1** *Nearest standard equivalents of some selected steels*

| *Steel type* | *BS970 (1970–83)* | *BS970 (1955) (En)* | *AISI/ SAE* | *Werk-stöff* | *Steel type* | *BS970 (1970–83)* | *BS970 (1955) (En)* | *AISI/ SAE* | *Werk-stöff* |
|---|---|---|---|---|---|---|---|---|---|
| CARBON | 040A10 | 2A | 1010 | 1.0301 | SILICO MANGANESE | 250A53 | 45 | 9255 | 1.0904 |
| | 070A72 | 42 | 1070 | – | | 250A58 | 45A | 9260 | 1.0909 |
| | 070M20 | 3A | 1020 | 1.0402 | STAINLESS HEAT RESISTING AND VALVE | 303S31 | 58M | 303 | 1.4305 |
| | 070M55 | 9 | 1055 | 1.0535 | | 303S42 | 58AM | 303Se | 1.4305 |
| | 080A15 | 3B | 1016 | – | | 304S15 | 58E | 304 | 1.4301 |
| | 080A32 | 5C | 1035 | 1.1180 | | 321S31 | – | 321 | 1.4541 |
| | 080A40 | 8C | 1040 | 1.1186 | | 403S17 | – | 403 | 1.4000 |
| | 080A47 | 43B | 1046 | 1.1730 | | 416S21 | 56AM | 416 | 1.4005 |
| | 080M15 | 32C | 1016 | – | | 416S37 | 56CM | – | – |
| | 080M30 | 6 | 1030 | 1.0560 | | 430S17 | 60 | 430 | 1.4016 |
| | 080M40 | 8 | 1040 | 1.1186 | | 431S29 | 57 | 431 | 1.4057 |
| | 080M50 | 43A | 1049 | – | ALLOY | 530A30 | 18A | 5130 | 1.7030 |
| CARBON MANGANESE | 150M19 | 14A | 1024 | 1.1160 | | 605M36 | 16 | – | – |
| | 150M36 | 15 | – | 1.1167 | | 635M15 | 351 | – | – |
| CARBON AND CARBON MANGANESE FREE CUTTING | 210M15 | 32M | 1117 | 1.0723 | | 655M13 | 36 | – | 1.5752 |
| | 212A42 | 8DM | – | – | | 665M17 | 34 | 4615 | – |
| | 212M36 | 8M | 1140 | – | | 708M40 | 19 | 4140 | 1.7225 |
| | 212M44 | 8M | – | – | | 709M40 | 19 | – | 1.7225 |
| | 214M15 | 202 | 1118 | – | | 735A50 | 47 | 6150 | 1.8159 |
| | 220M07 | 1A | 1113 | 1.0711 | | 817M40 | 24 | 4340 | 1.6565 |
| | 226M44 | 8M | 1144 | – | | | | | |
| | 230M07 | 1A | 1213 | 1.0715 | | | | | |

### 9.5.3 The classification of steels

Steels can be loosely classified by their composition and properties into six groups, namely:

(a) low carbon free-cutting steels
(b) low carbon steels
(c) medium carbon and carbon manganese steels
(d) alloy case-hardening steels
(e) alloy direct-hardening steels
(f) stainless steels

(a) **Low carbon free-cutting steels** are widely used for component parts machined from standard stock sections. Amongst the most popular are 220M07 and 230M07 (En1A); leaded versions for even better machining properties are also available. Where case-hardened surfaces are required 210M15 and 214M15 (En32M and En202) are semi-free-cutting possibilities.
(b) **Low carbon steels** are usually referred to as **mild steels** and are suitable for forming, bending, case-hardening and welding, for applications where they will be lightly stressed. The most commonly used include 040A10, 070M20, 080A15, 080M15 (En2A, En3A, En3B, En32C).
(c) **Medium carbon and carbon manganese steels** can be heat-treated by a variety of methods to give greater strength and improved wear resistance than the mild steels. They are specified for shafts, gears, studs, bolts, rollers and so on. Tensile strengths up to 1000 $N/mm^2$ can be obtained but the range 700–850 $N/mm^2$ is more usual. Choose from 080M30, 080M40, 080A40, 080M50, 070M55, 150M19 (En5, En8, En8C, En43A, En9, En14A) and, where a considerable amount of machining is required, consider the free-cutting versions 212M36, 212M44, 212A42, 226M44 (En8M and En8DM).
(d) **Alloy case-hardening steels** can provide hard wear-resisting surfaces for components subject to shock loading; they combine a hard exterior with a tough core. However, it is important to understand the significance of 'ruling section' (Section 9.5.4.4) and so to avoid thin cross-sections when using the more highly alloyed types; through-hardening may be unavoidable with the consequent brittleness due to the loss of the impact-resisting core. Important steels of this type include 635M15, 655M13, 665M17 (En33, En34, En36, En351).
(e) **Alloy direct hardening steels** are used where high strength and impact resistance are required. Alloying elements include nickel, chromium, molybdenum and vanadium in various combinations and proportions. Depending on their end use, components may be machined from softened material and then hardened and tempered to the required range of strengths or they may be machined directly from heat-treated bar stock. Representative types are 605M36, 708M40, 709M40, 817M40 (En16, En19, En24).
(f) **Stainless steels** may be martensitic, ferritic or austenitic – for an explanation of these definitions refer to a relevant textbook in engineering metallurgy.

*Martensitic stainless steels* can be hardened and tempered to 550–1000 $N/mm^2$ tensile strengths: they are used for fasteners, valves and shafts as well as surgical instruments and cutlery. Free-cutting types include 416S21 and 416S37 (En56AM, En56CM).

*Ferritic stainless steels* are ductile and so can be easily formed from strip or sheet for applications such as motor car trims and exhaust systems, food processing and catering equipment etc. Commonly used types are 403S17 and 430S17 (En60).

*Austenitic stainless steels* are the most resistant to corrosion and are used in the food, chemical, gas and oil industries as well as for domestic appliances, motor car parts, medical equipment and so on. They cannot be hardened by heat treatment but they do work harden (and so may require inter-stage annealing during manufacture of components). They are also non-magnetic. Free-cutting types include 303S31 and 303S42 (En58M). Others are 304S15 and 321S31 (En58B, En58C and En58E).

In addition to these six categories it is sometimes useful to identify *spring steels* 070A72 and 735A50 (En42 and En47) separately, due to the fact that they have very specific applications.

*Silver steel* is another useful material, mainly because it can be hardened and tempered relatively easily with the minimum of equipment. It is defined in BS1407 High Carbon Bright Steel.

Table 9.5.2 gives a useful summary of some of the mechanical properties of selected steels.

Where a range of values is given this generally indicates properties dependent upon heat treatment and ruling section. For those materials listed in Table 9.5.1 and not in Table 9.5.2, the relevant information was not available.

**Table 9.5.2** *Some mechanical properties of selected steels*

| Grade (BS970) | En | Tensile strength ($N/mm^2$) | Yield stress ($N/mm^2$) | Elongation (%) (after fracture) | Hardness (HB) |
|---|---|---|---|---|---|
| 070M20 | En3A | 430 | 215 | 21 | 126/179 |
| 070M55 | En9 | 700 | 355 | 12 | 201/255 |
| 080M15 | En32C | 350 | 175 | 22 | 109/163 |
| 080M30 | En6 | 490 | 245 | 20 | 143/192 |
| 080M40 | En8 | 625/850 | 385/465 | 16 | 179/255 |
| 080M50 | En43A | 620/570 | 295 | 14 | 163/217 |
| 150M19 | En14A | 550 | 325 | 18 | 152/207 |
| 150M36 | En15 | 620/600 | 355 | 14/15 | 179/223 |
| 210M15 | En32M | 490 | – | 16 | – |
| 212M36 | En8M | 550/850 | 340/495 | 20/16 | 152/255 |
| 214M15 | En202 | 740 | – | 12 | – |
| 220M07 | En1A | 360 | 215 | 22 | 103 |
| 226M44 | En8M | 700/1000 | 600 | 16/12 | 201/302 |
| 230M07 | En1A | 360 | 215 | 22 | 103 |
| 303S31 | En58M | 510 | – | 40 | 183 |
| 303S42 | En58AM | 510 | – | 40 | 183 |
| 304S15 | En58E | 480 | – | 40 | 183 |
| 321S31 | – | 510 | – | 35 | 183 |
| 403S17 | – | 420 | 280 | 20 | 170 |
| 416S21 | En56AM | 550/850 | 370/525 | 15/11 | 150/63 |
| 416S37 | En56CM | 700/925 | 525/585 | 11/10 | 201/277 |
| 430S17 | En60 | 430 | 280 | 20 | 170 |
| 431S29 | En57 | 850/1000 | 680 | 11 | 248/302 |
| 605M36 | En16 | 700/850 | 495 | 15 | 201/255 |
| 635M15 | En351 | 770 | – | 12 | – |
| 655M13 | En36A | 1000 | – | 9 | – |
| 665M17 | En34 | 770 | – | 12 | – |
| 708M40 | En19 | 700/1225 | 495/955 | 15/8 | 201/375 |
| 709M40 | En19 | 700/1225 | 495/955 | 15/8 | 201/375 |
| 735A50 | En47 | 850/1508 | 650/1250 | 13/3 | 248/444 |
| 817M40 | En24 | 850/1500 | 650/1235 | 12/5 | 248/444 |

### 9.5.4 Some definitions and explanations

#### 9.5.4.1 Crystal and grain structures

To understand the physical and manufacturing properties of steels and other alloys it is necessary to understand their structure. As steels solidify from the molten state, crystals are formed within each of which the atoms are arranged in an orderly manner according to a fixed pattern. If the rate of cooling at solidification is slow then relatively few crystals will be formed and these will grow to be relatively large; a coarse grain structure will result. With rapid cooling a larger number of crystals will start to grow but their size will be limited by the close proximity of their neighbours; a finer grain structure will result. Steels with larger grains are more ductile than those with smaller grains; thus they can be bent, or otherwise formed, to a greater extent but are more difficult to machine because the cutting tool tends to become embedded in the material.

#### 9.5.4.2 Heat-treatment

The mechanical, manufacturing and other properties of steels can be manipulated by subjecting them to particular cycles of heating and cooling, some of which are described below. Typically it may be desirable to make material harder or stronger or more ductile. Critical factors in heat treatment include temperature and rate of cooling (or 'quenching').

#### 9.5.4.3 Hardening

Hardness is a particularly desirable property to minimize wear when contacting surfaces move relative to one another and to minimize abrasion when surfaces work with abrasive materials, for example in earth moving, quarrying and mining.

Iron has a peculiar property in that the orderly arrangement of atoms within each crystal between 910°C and 1400°C is different from that below and above that range (face-centred cubic lattice rather than body-centred cubic lattice). With some other metallic alloys which exhibit a similar transformation it is possible to 'trap' a metallic structure as it existed at a higher temperature and preserve it at room temperature by cooling it rapidly (i.e. quenching it), but with iron and low carbon steels this is not possible because the speed of transformation is too fast. However, in steels with a carbon content greater than 0.3%, carbon atoms can be caught out of their room temperature position by quenching; the resultant distortion of the crystal structure on the molecular scale appears as hardness (and brittleness) of the steel component so treated.

The actual hardness achieved depends on the proportion of carbon present and the rate of cooling. The presence of other alloying elements in alloy steels also affects the end result; for example, alloy steels can be hardened with much less drastic quenching methods than those required with plain carbon steels.

#### 9.5.4.4 Ruling section

The hardness achieved by a particular heat treatment for a particular grade of steel depends upon the rate of cooling. The concept of a *ruling section* takes into account the effect of mass. Thin sections are easily cooled quickly but as size increases it becomes progressively difficult to cool the core of the section at the required rate: above a certain *ruling section* it is impossible to quench the steel fast enough to harden it completely to its central core. Ruling sections may be effectively increased by the addition of alloying elements to the steel; these reduce the critical rate of transformation and so make it possible to through-harden heavier sections.

#### 9.5.4.5 Hardness and toughness

Toughness describes a material's ability to resist impact or shock loading. Unfortunately, toughness is more or less inversely proportional to the hardness achieved by heat treatment: steels are more brittle after hardening.

#### 9.5.4.6 Tempering

Fully hardened steels can be made less brittle (but also less hard) by reheating to temperatures between 230°C and 400°C. The temperature and rates of heating and cooling chosen depend on the particular alloy involved and on the particular combination of hardness, toughness and ductility desired in the finished product.

When a clean, shiny steel surface is heated with, say, an oxyacetylene flame, the oxide film on the surface becomes visible initially as a light yellow colour. As the temperature rises, so the colour darkens through brown and purple to blue (but the colours do not reverse on cooling). For plain carbon steels each shade of colour represents a particular temperature; they can therefore be used for tempering small tools as indicated in Table 9.5.3.

**Table 9.5.3** *Temperatures associated with oxide film colouration in plain carbon steels*

| *Tempering temperature (°C)* | *Tempering colour* | *Typical components* |
|---|---|---|
| 220 | Pale straw | Scrapers, scribers, turning tools |
| 230 | Straw | Shaping/planing tools |
| 240 | Deep straw | Milling cutters, drills, reamers, boring tools |
| 250 | Light brown | Knife blades, taps, dies, woodwork tools |
| 260 | Purple-brown | Rivet snaps, punch tools |
| 270 | Purple | Axes, press tools, surgical tools |
| 280 | Deep purple | Steel cold chisels, wood chisels, soft wood planes |
| 290 | Bright blue | Wrought iron cold chisels, screwdrivers |
| 300 | Dark blue | Springs, joinery saws |

### *9.5.4.7 Case hardening*

The surface of low carbon steel components can be hardened by introducing more carbon atoms into the surface layers of the material by heating it with carbonaceous material to between 900°C and 950°C and holding the temperature for up to 5 hours depending on the depth of penetration required. When put through an appropriate heat treatment cycle, the modified surface steel will harden while the low carbon core will not, but rather will retain its original tough properties.

Nitriding is another process giving a hard surface to a tougher core material. Particular alloy steels have to be used, however, because the process relies on the formation of hard nitrides of metals such as aluminium, chromium and vanadium when the component part is exposed to atomic nitrogen at about 500°C for between 40 and 100 hours. No subsequent heat treatment involving quenching is required. Typical steels are 708M40 and 709M40.

### *9.5.4.8 Stress and strain*

When a section of material is subjected to a load, whether tensile, compressive or bending, it is said to be *stressed*. *Stress* is measured in terms of the size of the load carried by a unit area of the section. Materials change their size when they are loaded. *Strain* is a measure of the change in size (length) compared with the original (unstressed) length.

### *9.5.4.9 Strain hardening (work hardening)*

When a metal is stressed beyond its elastic limit while below its recrystallization temperature, plastic deformation takes place. As deformation proceeds, the metal becomes harder and increasingly resistant to further deformation until it reaches a stage where no further deformation is possible; the material is strain- or work-hardened by cold-work (Section 9.5.4.12).

If the applied stress is further increased, fracture will occur. Thus there is a definite limit to the extent to which a material's geometry may be changed by cold deformation (e.g. bending, stretching, hammering) below its recrystallization temperature.

### *9.5.4.10 Annealing and recrystallization*

Work-hardened materials may be returned to approximately their original properties of strength, hardness and ductility by a heat-treatment process called annealing; they can then be further deformed (cold-worked) if required.

In the initial stages of annealing at relatively low temperatures the atoms within the distorted crystal structure of the deformed material are able to rearrange themselves into less stressed positions. If the temperature is progressively raised, a point is reached where new crystals begin to grow from points within the deformed lattice. Given time, growth will continue until the new unstressed crystals completely absorb the deformed crystals. If left at the elevated temperature beyond this stage, certain crystals will continue to grow at the expense of the others; a coarse grain structure will result.

Each alloy has its own range of recrystallization temperature which is usually between one-third and one-half of its absolute melting point temperature (K). The actual recrystallization temperature in any instance depends on the degree of deformation present: the greater the degree of cold-work, the lower the recrystallization temperature. For some low melting point metals, such as lead and tin, their recrystallization temperature is below normal room temperature so they cannot be work-hardened by deforming since annealing is taking place whilst the deformation is in progress; they are in fact being hot-worked at normal room temperature.

#### *9.5.4.11 Normalizing*

Normalizing is in effect a variant on annealing. Instead of allowing the heated metal to cool slowly to ambient temperature, it is removed from the furnace and left to cool in still air. Grain growth is therefore curtailed and the mechanical properties resulting are somewhat better than those of the annealed equivalent, although it follows that the material's capacity for further cold-work (deformation) is restricted.

#### *9.5.4.12 Hot and cold working*

Hot-working processes are those like forming and forging where a metal or alloy is deformed plastically above its recrystallization temperature. The recrystallization temperatures of steels are generally within the range 450–650°C.

Provided steels shaped above their particular recrystallisation temperature are not cooled too soon or too quickly, the crystal structure will not remain deformed and the product will be ductile but usually difficult to machine. They will therefore be in a suitable condition to be the raw material for fabrication and welding. If, on the other hand, the product is forged as near as possible to the finished component (e.g. a connecting rod) then the temperature of the final stages will be carefully restricted to ensure that recrystallization is curtailed by early cooling. Thus the final structure achieved will comprise fine stress-free grains which have emerged from the crystals distorted by forging but have not had time to grow to a coarser structure with associated inferior mechanical performance.

Cold-working deformation processes are those carried out below a metal's recrystallization temperature. The crystal structure is deformed by the process and only a limited change in shape is possible before cracking or other failure occurs. For larger changes in shape the material has to be annealed, to allow the crystals to regroup into an unstressed condition. The crystal structure of cold-worked products remains distorted. The material is usually stronger and harder than its hot-worked equivalent but is likely to be more brittle; if subsequently bent or otherwise deformed, it will fail much sooner.

#### *9.5.4.13 Bright and black bar stock*

Black bars – whether round, square, flat, angle, channel or 'I' section – are rolled to shape whilst red-hot, at temperatures as high as 1200°C. At such temperatures, steel forms scale as it interacts with atmospheric gases. This scale is hard and abrasive and liable to flake off during the production process; hence some of it may be rolled temporarily into the surface of the metal. Thus, although the product is sold descaled, it is not only black in appearance but there are pits and other irregularities left on the surface. Generally speaking, black bar stock is bought in an annealed or hot-worked condition and so does not machine well. It is, however, comparatively stress-free and ideal for welding and other fabrication purposes.

Bright bars are often referred to as 'cold-rolled' but a more accurate description would be cold-finished; although strip and sheet steel is cold-rolled, other sections are finished by drawing them through carbide dies, i.e. they are 'cold-drawn'.

In the main stages of reduction, bright bars are rolled in the same way as black bars but before the finishing cold drawing stages they are descaled. The subsequent drawing processes compress the surface, making it smooth and shiny. The finished bars are also close to their nominal size, typically within 0.125 mm. They exhibit the properties of cold-worked steel, having increased tensile and yield strengths, reduced ductility and improved machining properties. Some round bright bars may be machined by turning or grinding as finishing operations prior to sale.

It should be noted that residual stresses are inherent in bright section stock. So if material is removed by machining, the stresses left will adopt a new state of equilibrium which is likely to result in distortion of the component if it is not symmetrical.

#### *9.5.4.14 Machinability and free-cutting steels*

The term machinability defies an exact definition because, while expressing the relative ease with which a particular material may be machined, the criteria by which it is assessed varies with different applications. For example, maximum metal removal rate may be important in one instance while a good surface finish or ease of swarf disposal may be required in others.

In all machining processes, the material removed from the work piece as chips or swarf is caused to fail by shear stress on the shear plane. As in any highly stressed area, discontinuities in geometry (e.g. sharp internal corners) or in material structure (e.g. slag inclusions in the crystalline structure) are likely to become stress-raising or stress-concentrating features from which cracks propagate, causing failure at a level of loading or stress much below that which would be expected if the discontinuity did not exist: the stress-raising effect of the discontinuities can be out of all proportion to their size. In free-cutting materials, widely dispersed minute discontinuities are introduced into the material's structure such that they allow low stress failure on the shear plane but have little effect on the material's general properties.

In mild steels the sulphur content is normally minimized, to below 0.06%, because it forms brittle inter-crystalline films of ferrous sulphide, thus weakening the steel. If, however, 0.9–1.2% manganese is added to a steel with up to 0.2% sulphur content, the manganese combines preferentially with the sulphur to form manganese sulphide which exists as isolated globules in the micro-structure. The resulting steel thus has free machining characteristics whilst maintaining good mechanical properties. Other sulphide-forming elements used in more sophisticated free-cutting steels include molybdenum, zirconium and selenium.

Lead is an element which does not dissolve in steels or in a variety of non-ferrous alloys. When amounts between 0.15 and 0.35% are added to steels (carbon and stainless), it disperses as globules so small that they cannot be detected with a simple microscope; however, machinability is improved by up to 50%.

Free-cutting steels may be marginally more expensive than their untreated equivalents but the improvement in machinability more than pays for the additional charge in all but the very simplest machining applications.

Table 9.5.4 gives the relative machinability of a number of major steels using the machinability of 230M07 (En1A) set at a value of 100. Note that the higher the value the better the machinability.

**Table 9.5.4** *Machinability figures for selected steels*

| *BS970 (1983)* | *BS970 (1955)* | *Index* |
|---|---|---|
| **230M07** | **En1A** | **100** |
| 230M07 Pb | En1A | 150 |
| 230M07 Pb Te | En1A Pb Te | 182 |
| 220M07 | En1A | 95 |
| 070M07 | En3A | 55 |
| 080M15 | En32B | 55 |
| 210M15 | En32M | 77 |
| 210M15 Pb | En32M Pb | 100 |
| 214M15 | En202 | 80 |
| 080M40 | En8 | 45 |
| 080M40 Pb | En8 Pb | 58 |
| 212M36 | En8M | 58 |
| 212M36 Pb | En8M Pb | 76 |
| 070M55 | En9 | 30 |
| 150M36 | En15R | 47 |
| 212M36 | En15AMR | 47 |
| 212M36 Pb | En15AMR Pb | 61 |
| 605M36 | En16T | 29 |
| 605M36 Pb | En16T Pb | 38 |
| 708M40 | En19T | 29 |
| 817M40 | En24T | 29 |

### 9.5.4.15 Weldability

The weldability of steels depends on their composition; in particular their carbon content is a key factor. Low carbon steels with up to 0.25% are readily welded but as carbon and/or other alloying constituents increase so the likelihood of cracking occurring in the heat-affected zone also increases. In assessing the susceptibility of a steel to this form of cracking the carbon equivalent (CE) formula is used (the lower the value the better):

$$CE = C + \frac{Mn}{6} + \frac{Cr + Mo + V}{5} + \frac{Ni + Cu}{15}$$

Where

C = percentage carbon
Mn = percentage manganese
Cr = percentage chromium
Mo = percentage molybdenum
V = percentage vanadium
Ni = percentage nickel
Cu = percentage copper

Special precautions and appropriate filler materials may be required for certain of these alloy steels; welding rod manufacturers give recommendations for each particular standard grade of steel.

# 10
# Design for manufacture

## 10.1 Limits and fits and process capabilities

### *10.1.1 Selected ISO fits – hole basis (see BS4500 for more details)*

Some of the principal terms relating to limits and fits are given in the figure. Although only cylindrical parts are represented as holes and shafts it must be remembered that the terms should be applied in their broadest sense.

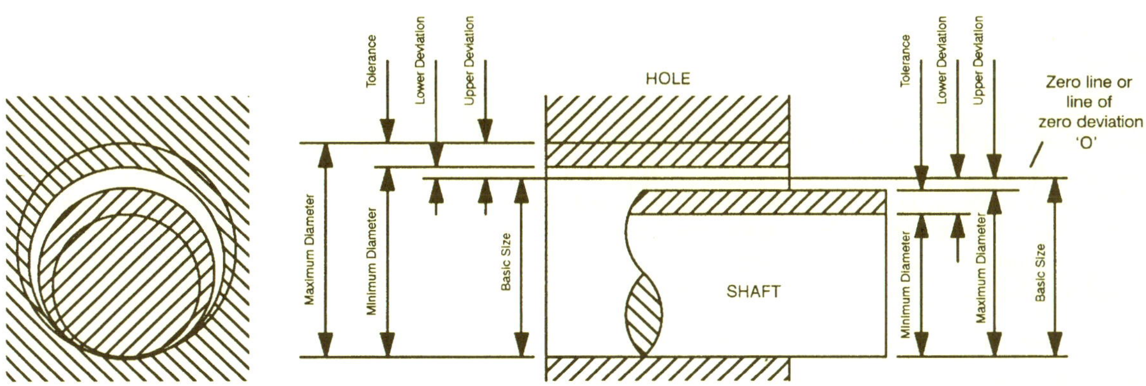


For most general applications the hole-basis system of fits is preferred to the shaft-basis system: the size of the hole is fixed as a datum and the shaft size is selected to give the required fit (for the shaft-basis the opposite is the case, as shown in the next figure).

The principal reason for preferring the hole-basis system is that the size of holes (particularly small ones less than 20 mm) is usually dependent on the tools used to make them, e.g. drills, reamers, etc., whereas the size of most shafts is achieved by the machine tool used in its making, rather than relying on the cutting tool. Also, small holes are usually checked with standard gauges – the fewer the hole sizes, the fewer the gauges required; shafts are of course measured with adjustable instruments (e.g. micrometers, vernier callipers).

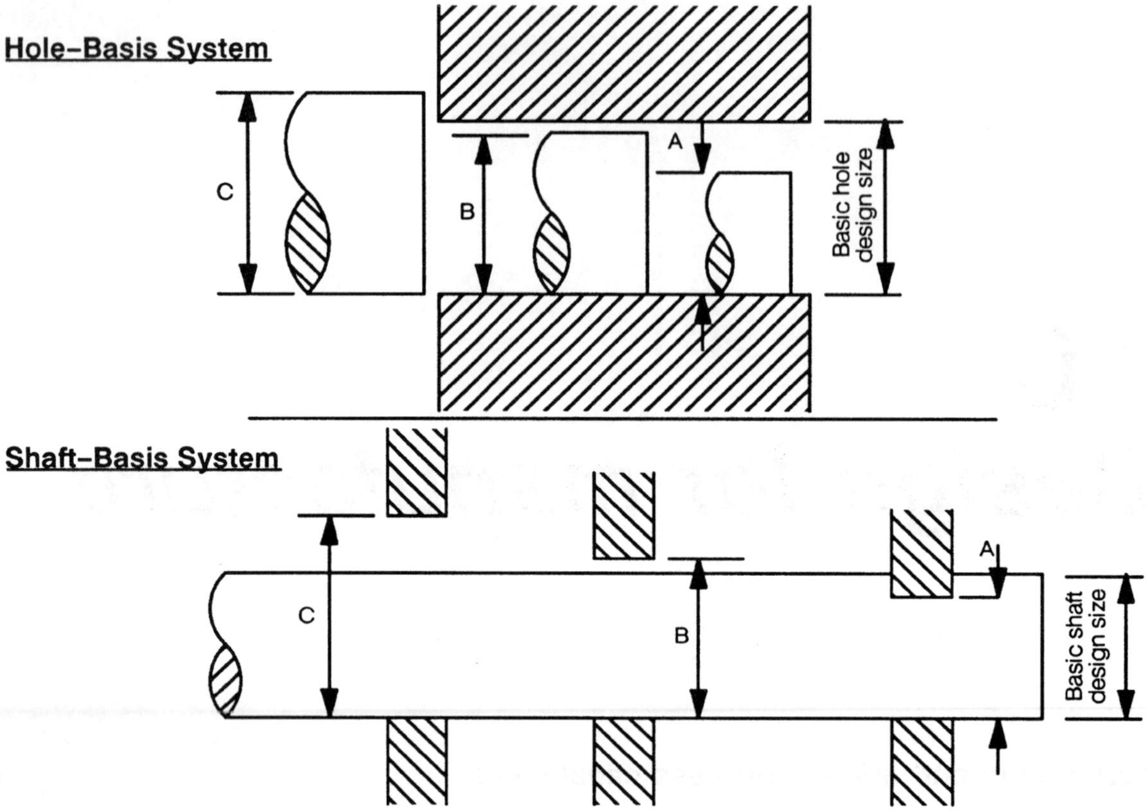


In the British Standard (BS4500), shown on p. 165, the whole range of possible fits is divided into three main classes:

(a) clearance
(b) transition
(c) interference.

Each main class is then further divided into several types of fit; for example, clearance fits are divided into six types from very fine to very coarse, each type of fit being classified by letter symbols.

The wide range of possible fits which the Standard allows may be seen as a strength in that it can cope with virtually every conceivable application. In practice the Standard recommends selected fits to satisfy general requirements (e.g. loose running fits, loose clearance fits, normal running fits, precision sliding fits – similar requirements are identified for transition and interference fits).

In using this tabulation of limits and fits, note that the horizontal line marked 'O' in the shaft/hole diagram represents the exact (nominal) size of the hole or shaft. Hatched areas above this line represent tolerance zones for sizes greater than the nominal and hatched areas below the line represent tolerance zones for sizes less than nominal size.

The selected fits allow for six possible clearance fits, two transition fits and two interference fits. They are based on four different tolerances (H11, H9, H8 and H7) on the nominal hole size which, at its smallest, is never less (defined by the letter H) than the nominal size; the actual allowable variation or tolerance is defined by the accompanying number (11, 9, 8 or 7).

The position of the tolerances on the shaft size relative to the nominal size are indicated by the letters c, d, e, f, g, h, k, n, p and s while the extent of the tolerance is represented by the numbers 11, 10, 9, 7 and 6. Thus, for example, H and h both represent tolerances for holes and shafts respectively which, at their maximum material condition, would give exact features. Similarly, 11 represents a particular tolerance for any particular size: for a 10 mm hole or shaft, 11 represents a tolerance of 0.110 mm.

# SELECTED ISO FITS – HOLE BASIS

(Data extracted from BS 4500A – 1985)

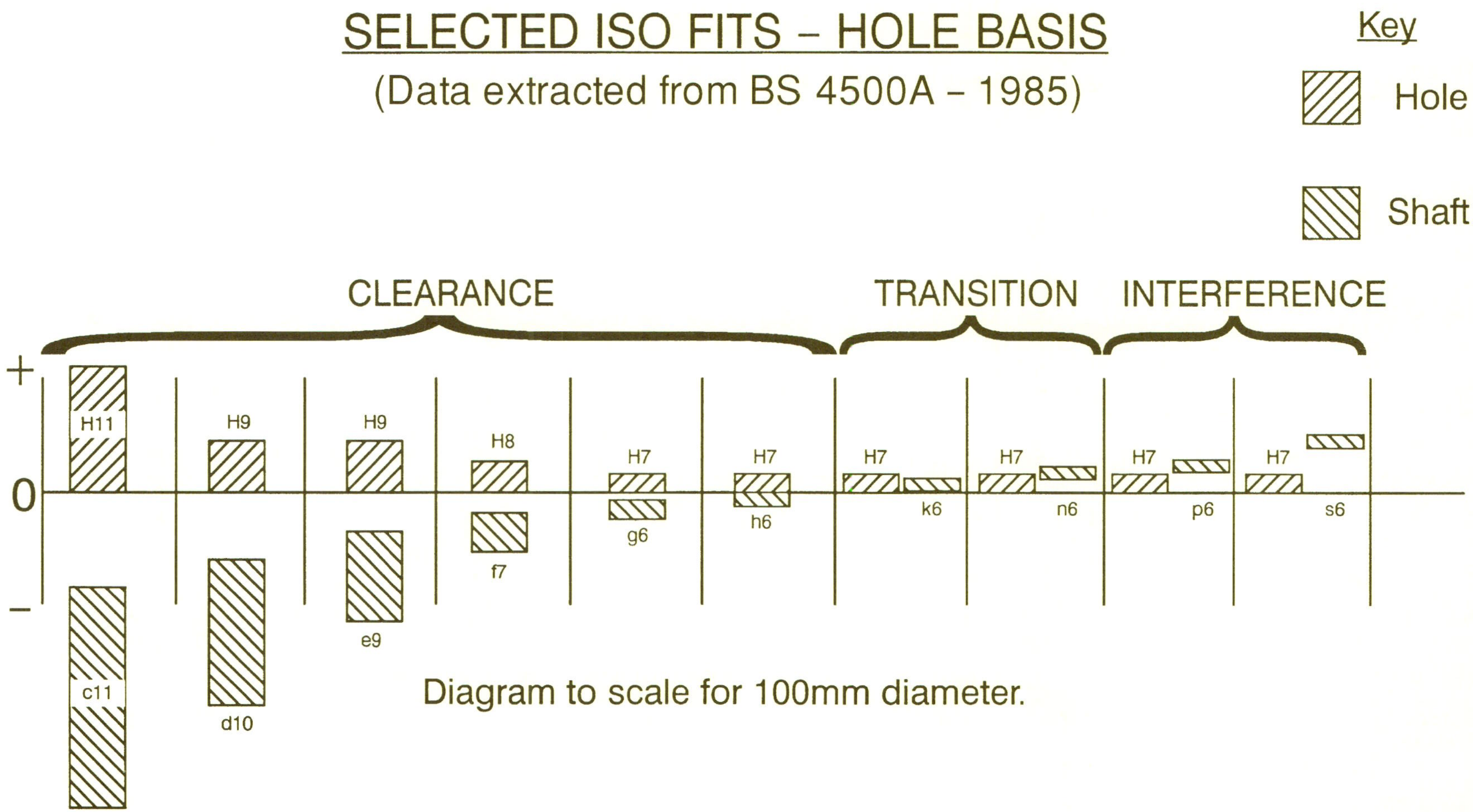


Diagram to scale for 100mm diameter.

| Nominal sizes | | Tolerance | | Tolerance | | Tolerance | | Tolerance | | Tolerance | | Tolerance | | Tolerance | | Tolerance | | Tolerance | | Tolerance | | Nominal sizes | |
|---|---|---|---|---|---|---|---|---|---|---|---|---|---|---|---|---|---|---|---|---|---|---|---|
| Over | To | H11 | c11 | H9 | d10 | H9 | e9 | H8 | f7 | H7 | g6 | H7 | h6 | H7 | k6 | H7 | n6 | H7 | p6 | H7 | s6 | Over | To |
| mm | mm | 0.001 mm | 0.001 mm | 0.001 mm | 0.001 mm | 0.001 mm | 0.001 mm | 0.001 mm | 0.001 mm | 0.001 mm | 0.001 mm | 0.001 mm | 0.001 mm | 0.001 mm | 0.001 mm | 0.001 mm | 0.001 mm | 0.001 mm | 0.001 mm | 0.001 mm | 0.001 mm | mm | mm |
| – | 3 | +60<br>0 | −60<br>−120 | +25<br>0 | −20<br>−60 | +25<br>0 | −14<br>−39 | +14<br>0 | −6<br>−16 | +10<br>0 | −2<br>−8 | +10<br>0 | −6<br>0 | −10<br>0 | +6<br>0 | +10<br>0 | +10<br>+4 | +10<br>0 | +12<br>+6 | +10<br>0 | +20<br>+14 | – | 3 |
| 3 | 6 | +75<br>0 | −70<br>−145 | +30<br>0 | −30<br>−78 | +30<br>0 | −20<br>−50 | +18<br>0 | −10<br>−22 | +12<br>0 | −4<br>−12 | +12<br>0 | −8<br>0 | +12<br>0 | +9<br>+1 | +12<br>0 | +16<br>+8 | +12<br>0 | +20<br>+12 | +12<br>0 | +27<br>+19 | 3 | 6 |
| 6 | 10 | +90<br>0 | −80<br>−170 | +36<br>0 | −40<br>−98 | +36<br>0 | −25<br>−61 | +22<br>0 | −13<br>−28 | +15<br>0 | −5<br>−14 | +15<br>0 | −5<br>0 | +15<br>0 | +10<br>+1 | +15<br>0 | +19<br>+10 | +15<br>0 | +24<br>+15 | +15<br>0 | +32<br>+23 | 6 | 10 |
| 10 | 18 | +110<br>0 | −95<br>−205 | +43<br>0 | −50<br>−120 | +43<br>0 | −32<br>−75 | +27<br>0 | −16<br>−34 | +18<br>0 | −6<br>−17 | +18<br>0 | −11<br>0 | +18<br>0 | +12<br>+1 | +18<br>0 | +23<br>+12 | +18<br>0 | +29<br>+18 | +18<br>0 | +39<br>+28 | 10 | 18 |
| 18 | 30 | +130<br>0 | −110<br>−240 | +52<br>0 | −65<br>−149 | +52<br>0 | −40<br>−92 | +33<br>0 | −20<br>−41 | +21<br>0 | −7<br>−20 | +21<br>0 | −13<br>0 | +21<br>0 | +15<br>+2 | +21<br>0 | +28<br>+15 | +21<br>0 | +35<br>+22 | +21<br>0 | +48<br>+35 | 18 | 30 |
| 30 | 40 | +160<br>0 | −120<br>−280 | +62<br>0 | −80<br>−180 | +62<br>0 | −50<br>−112 | +39<br>0 | −25<br>−50 | +25<br>0 | −9<br>−25 | +25<br>0 | −16<br>0 | +25<br>0 | +18<br>+2 | +25<br>0 | +33<br>+17 | +25<br>0 | +42<br>+26 | +25<br>0 | +59<br>+43 | 30 | 40 |
| 40 | 50 | +160<br>0 | −130<br>−290 | | | | | | | | | | | | | | | | | | | 40 | 50 |
| 50 | 65 | +190<br>0 | −140<br>−330 | +74<br>0 | −100<br>−220 | +74<br>0 | −60<br>−134 | +46<br>0 | −30<br>−60 | +30<br>0 | −10<br>−29 | +30<br>0 | −19<br>0 | +30<br>0 | +21<br>+2 | +30<br>0 | +39<br>+20 | +30<br>0 | +51<br>+32 | +30<br>0 | +72<br>+53 | 50 | 65 |
| 65 | 80 | +190<br>0 | −150<br>−340 | | | | | | | | | | | | | | | | | +30<br>0 | +78<br>+59 | 65 | 80 |
| 80 | 100 | +220<br>0 | −170<br>−190 | +87<br>0 | −120<br>−260 | +87<br>0 | −72<br>−159 | +54<br>0 | −36<br>−71 | +35<br>0 | −12<br>−34 | +35<br>0 | −22<br>0 | +35<br>0 | +25<br>+3 | +35<br>0 | +45<br>+23 | +35<br>0 | +59<br>+37 | +35<br>0 | +93<br>+71 | 80 | 100 |
| 100 | 120 | +220<br>0 | −180<br>−400 | | | | | | | | | | | | | | | | | +35<br>0 | +101<br>+79 | 100 | 120 |

| Nominal sizes | | Tolerance | | Tolerance | | Tolerance | | Tolerance | | Tolerance | | Tolerance | | Tolerance | | Tolerance | | Tolerance | | Tolerance | | Nominal sizes | |
|---|---|---|---|---|---|---|---|---|---|---|---|---|---|---|---|---|---|---|---|---|---|---|---|
| Over | To | H11 | c11 | H9 | d10 | H9 | e9 | H8 | f7 | H7 | g6 | H7 | h6 | H7 | k6 | H7 | n6 | H7 | p6 | H7 | s6 | Over | To |
| mm | mm | 0.001 mm | 0.001 mm | 0.001 mm | 0.001 mm | 0.001 mm | 0.001 mm | 0.001 mm | 0.001 mm | 0.001 mm | 0.001 mm | 0.001 mm | 0.001 mm | 0.001 mm | 0.001 mm | 0.001 mm | 0.001 mm | 0.001 mm | 0.001 mm | 0.001 mm | 0.001 mm | mm | mm |
| 120 | 140 | +250<br>0 | −200<br>−450 | | | | | | | | | | | | | | | | | +40<br>0 | +117<br>+92 | 120 | 140 |
| 140 | 160 | +250<br>0 | −210<br>−460 | +100<br>0 | −145<br>−305 | +100<br>0 | −84<br>−185 | +63<br>0 | −43<br>−83 | +40<br>0 | −14<br>−39 | +40<br>0 | −25<br>0 | +40<br>0 | +28<br>+3 | +40<br>0 | +52<br>+27 | +40<br>0 | +68<br>+43 | +40<br>0 | +125<br>+100 | 140 | 160 |
| 160 | 180 | +250<br>0 | −230<br>−480 | | | | | | | | | | | | | | | | | +40<br>0 | +133<br>+108 | 160 | 180 |
| 180 | 200 | +290<br>0 | −240<br>−530 | | | | | | | | | | | | | | | | | +46<br>0 | +151<br>+122 | 180 | 200 |
| 200 | 225 | +290<br>0 | −260<br>−550 | +115<br>0 | −170<br>−355 | +115<br>0 | −100<br>−215 | +72<br>0 | −50<br>−96 | +46<br>0 | −15<br>−44 | +46<br>0 | −29<br>0 | +46<br>0 | +33<br>+4 | +46<br>0 | +60<br>+31 | +46<br>0 | +79<br>+50 | +46<br>0 | +159<br>+130 | 200 | 225 |
| 225 | 250 | +290<br>0 | −280<br>−570 | | | | | | | | | | | | | | | | | +46<br>0 | +169<br>+140 | 225 | 250 |
| 250 | 280 | +320<br>0 | −300<br>−620 | +130<br>0 | −190<br>−400 | +130<br>0 | −110<br>−240 | +81<br>0 | −56<br>−108 | +52<br>0 | −17<br>−49 | +52<br>0 | −32<br>0 | +52<br>0 | +36<br>+4 | +52<br>0 | +66<br>+34 | +52<br>0 | +88<br>+56 | +52<br>0 | +190<br>+158 | 250 | 280 |
| 280 | 315 | +320<br>0 | −330<br>−650 | | | | | | | | | | | | | | | | | +52<br>0 | +202<br>+170 | 280 | 315 |
| 315 | 355 | +360<br>0 | −360<br>−720 | +140<br>0 | −210<br>−440 | +140<br>0 | −125<br>−265 | +89<br>0 | −62<br>−119 | +57<br>0 | −18<br>−54 | +57<br>0 | −36<br>0 | +57<br>0 | +40<br>+4 | +57<br>0 | +73<br>+17 | +57<br>0 | +98<br>+62 | +57<br>0 | +226<br>+190 | 315 | 355 |
| 355 | 400 | +360<br>0 | −400<br>−760 | | | | | | | | | | | | | | | | | +57<br>0 | +244<br>+208 | 355 | 400 |
| 400 | 450 | +400<br>0 | −440<br>−840 | +155<br>0 | −230<br>−480 | +155<br>0 | −135<br>−290 | +97<br>0 | −68<br>−131 | +63<br>0 | −20<br>−60 | +63<br>0 | −40<br>0 | +63<br>0 | +45<br>+5 | +63<br>0 | +80<br>+40 | −63<br>0 | +108<br>+68 | +63<br>0 | +272<br>+232 | 400 | 450 |
| 450 | 500 | +400<br>0 | −480<br>−880 | | | | | | | | | | | | | | | | | +63<br>0 | +292<br>+252 | 450 | 500 |

### *10.1.2 Ascribing tolerances to dimensions*

The tolerance on a size can be specified as a deviation in one direction only from the nominal size (unilateral) or in either direction from the nominal (bilateral), as shown in the figure.

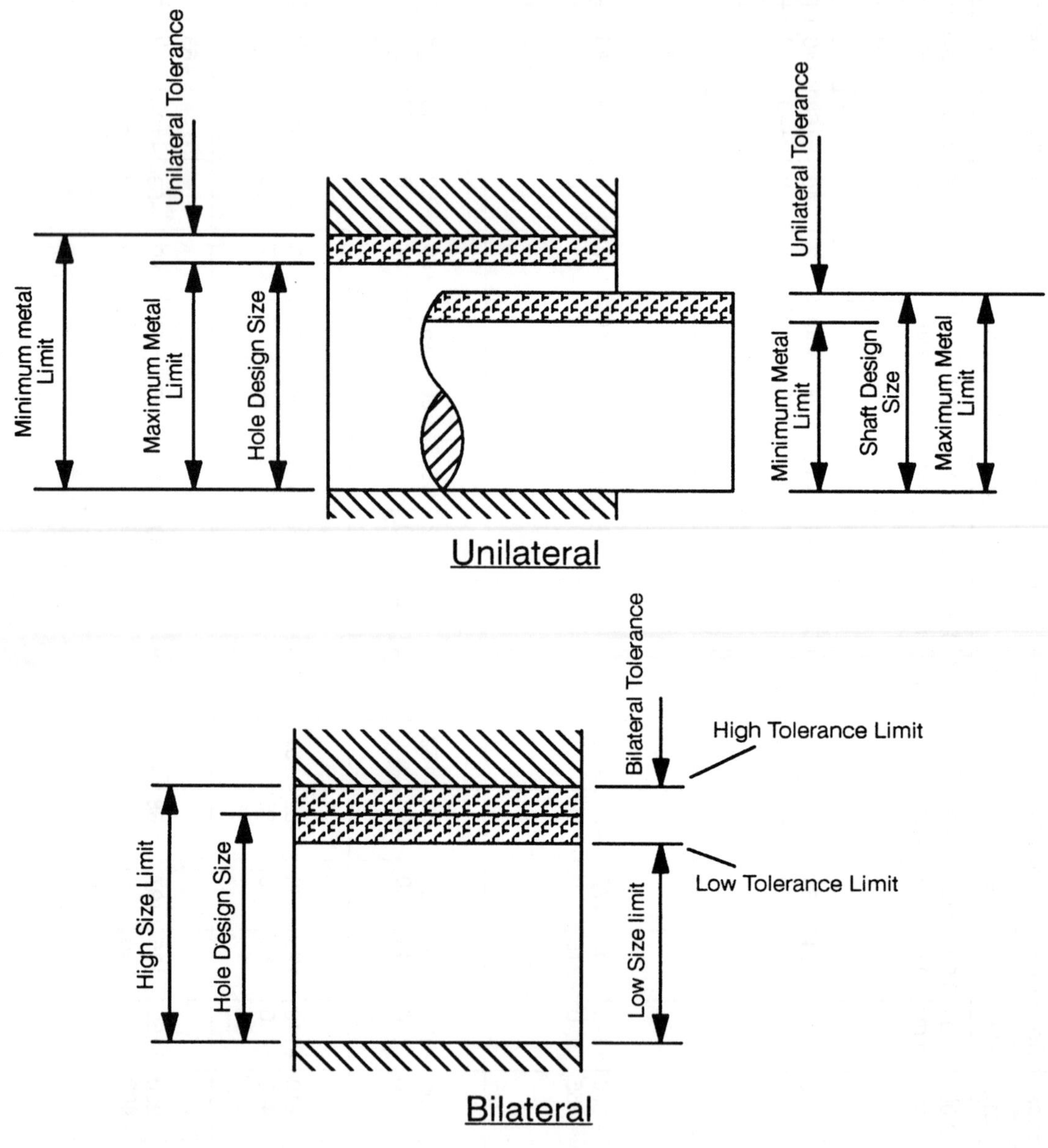


Unilateral tolerances are arranged so that the nominal size gives the tightest fit, automatically giving a situation in which the nominal size is the 'maximum material condition'. Since this means that further material could be removed without the dimension necessarily going outwith the tolerance, this arrangement generally results in less scrap because the tendency is for manufacture to aim at the nominal size.

### *10.1.3 Dimensioning*

Although *BS308 Engineering Drawing Practice, Part 2–Dimensioning and Tolerancing of Size* and other publications give valuable guidance in correct procedure, dimensioning remains something of an art. Care must be taken if contradictions and unacceptable accumulations of tolerances are to be avoided. For example, the following illustrations show how different dimensioning schemes allow different results. In the first instance, applying general tolerances to all three dimensions produces the contradictory result that the overall dimension falls outwith its specified tolerances if the other two dimensions are at their upper limit.

Removing one dimension (or designating it as an auxiliary dimension by the use of brackets) avoids the contradiction but components produced to the second scheme will not necessarily fall within the acceptable limits of the third scheme (and vice versa).

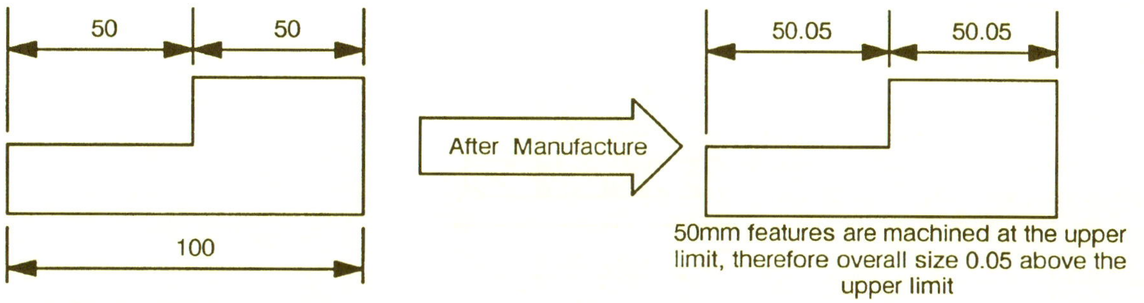


(i) General Tolerance ±0.05

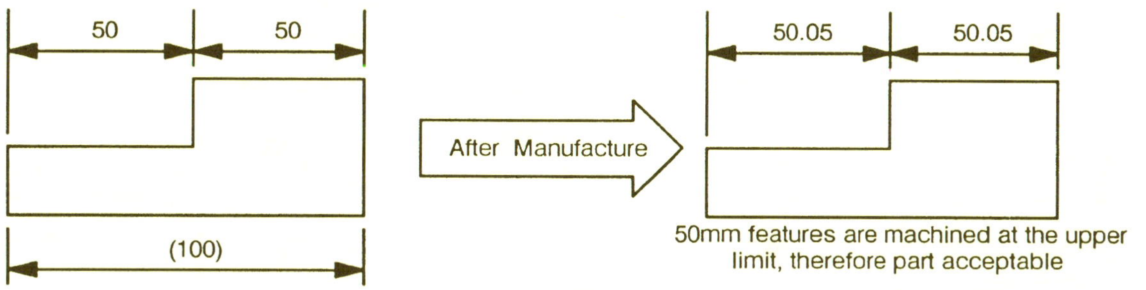


(ii) General Tolerance ±0.05

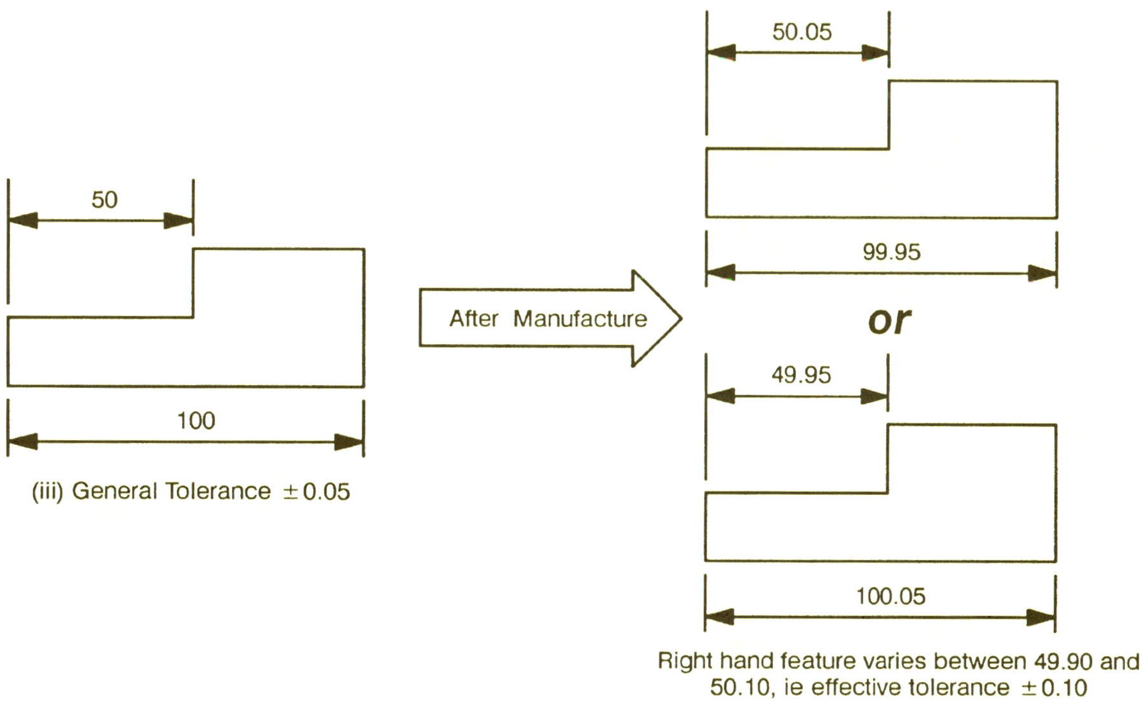


(iii) General Tolerance ±0.05

In applying a dimensioning scheme, as in the next figure, functionally important features of the component should be chosen as datums (although it should be emphasized that a functional datum may not necessarily be an important manufacturing feature): a good understanding of the function of the component is therefore a pre-requisite. The illustrations in the figure identify the functional dimensions of the components which contribute to the end float of a lever on its pivot pin and then show two different dimensioning schemes, only one of which controls the functional dimensions correctly. This scheme allows the tolerances on critical sizes to be 2.5 times as large (on average) as those dimensioned by the other; machining will be easier and, therefore, cheaper.

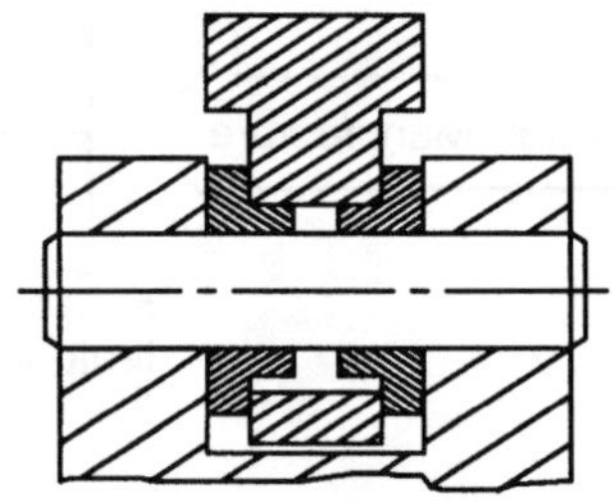

Functional dimensions governing the end float of the trunion bearing are marked ☞
(NB This does not mean all marked dimensions are the same size).

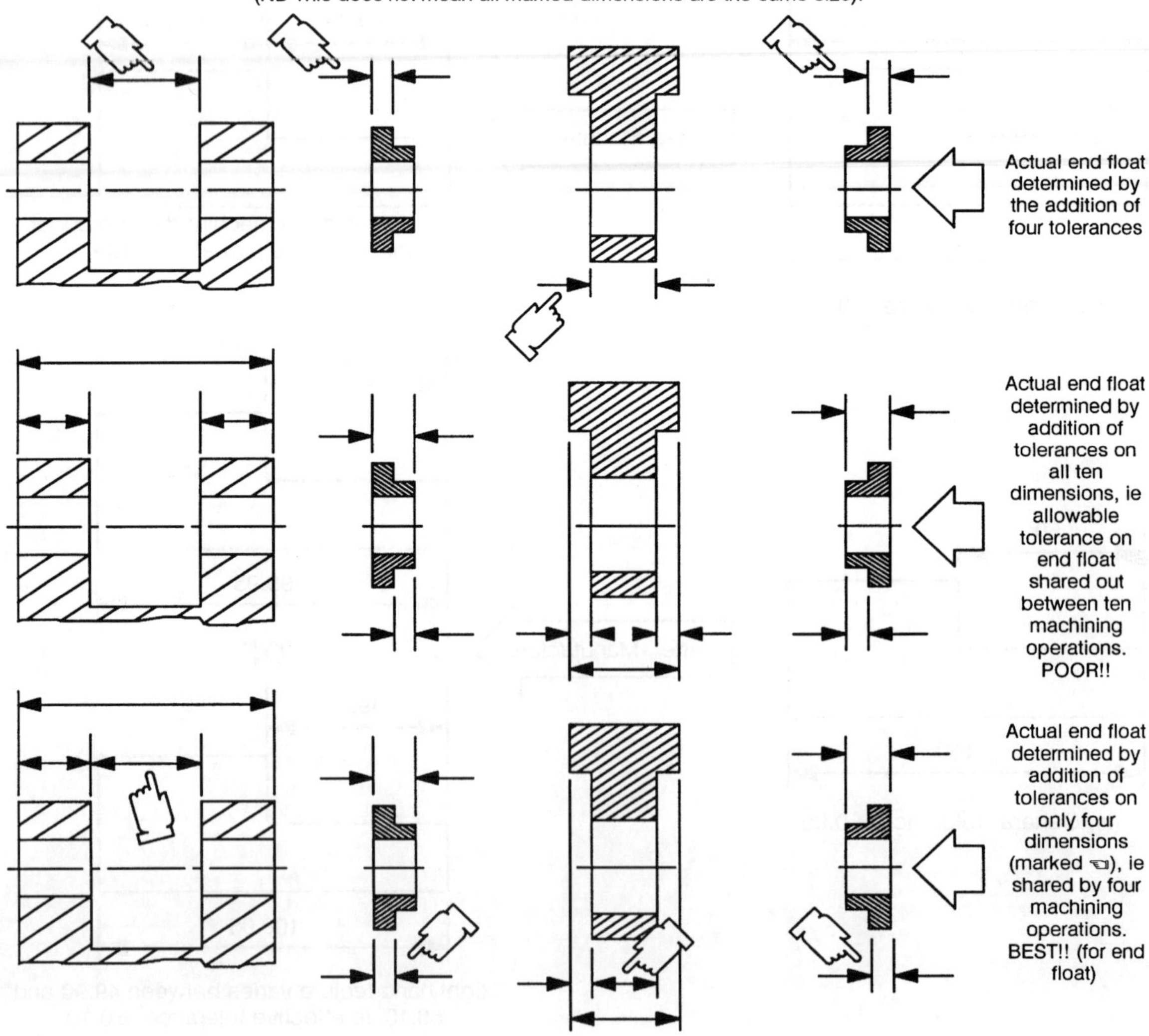

The next figure shows that there is a definite relationship between dimensional tolerance and surface finish tolerance. That is, for any particular process, a certain dimensional tolerance will be accompanied by a corresponding level of surface roughness. However the converse is unlikely to be true.

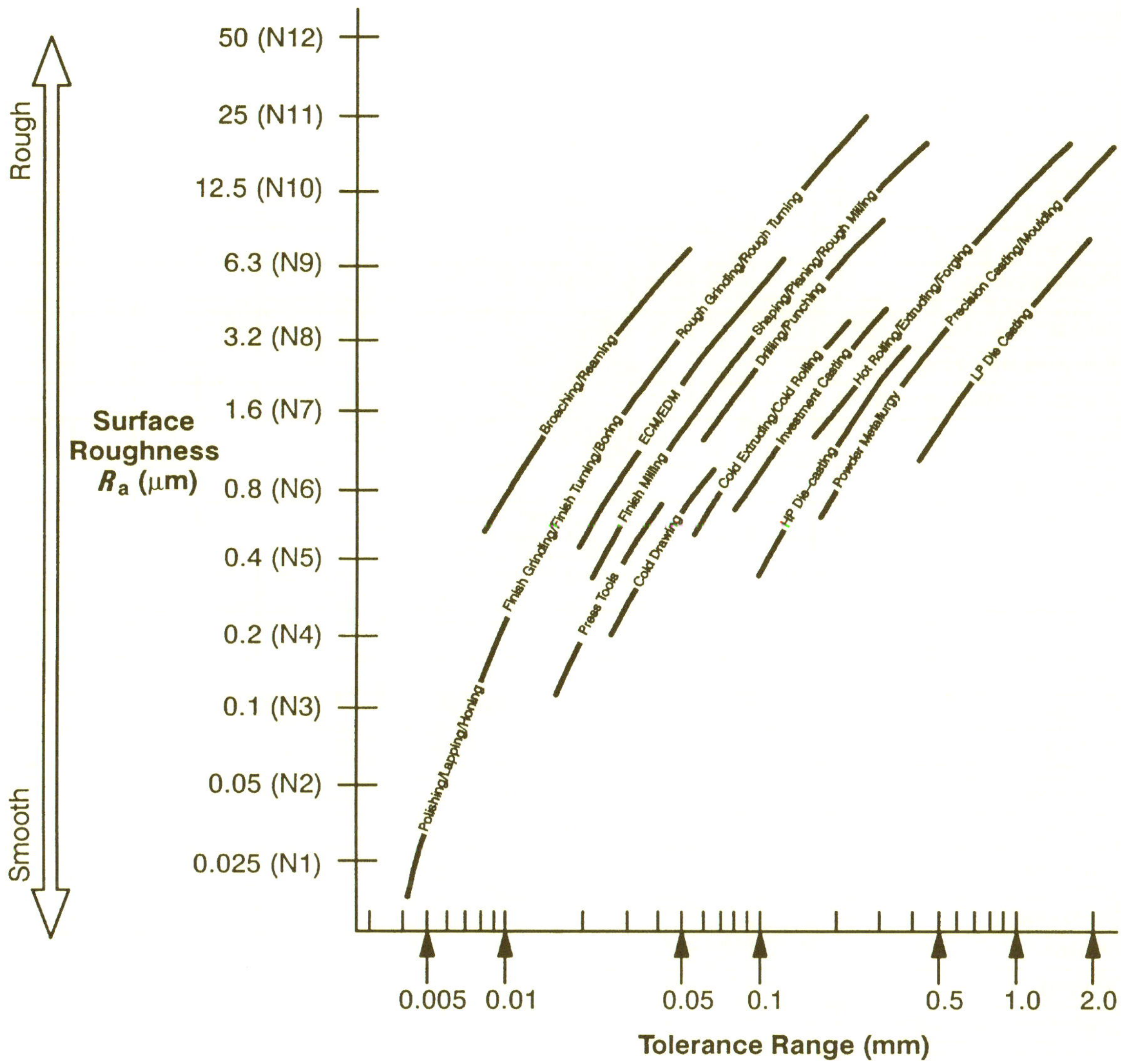

The following figure shows how the surface roughness is dependent on the manufacturing process used. Using this, the preferred process for producing a specific roughness can be found. However, should the need arise to use a non-preferred, though probably more expensive, process then the appropriate roughness values are also given.

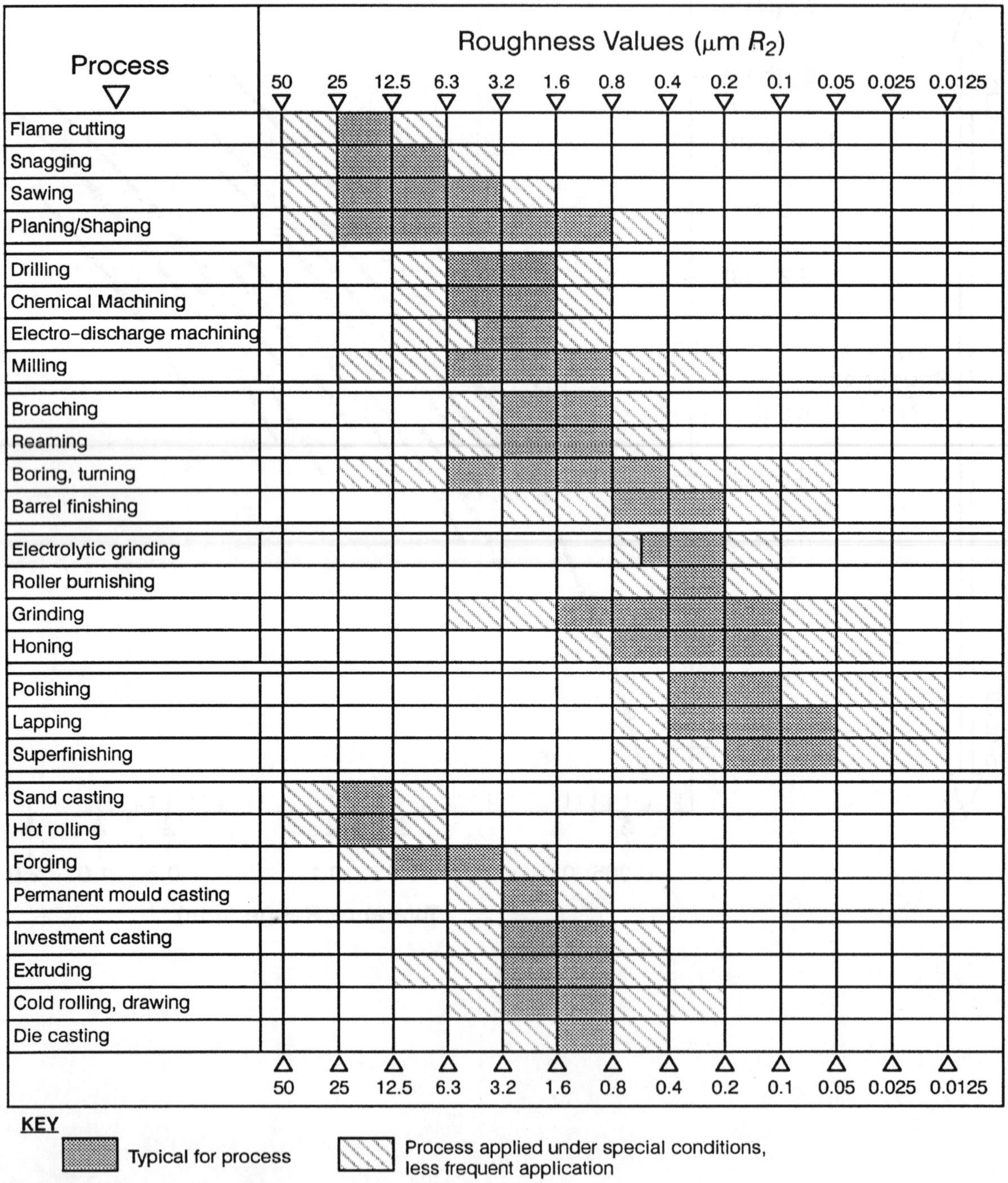


(from BS 1134: Part 2, 1990)

The final figure in this section gives a relative comparison of processes for specific tolerance ranges.

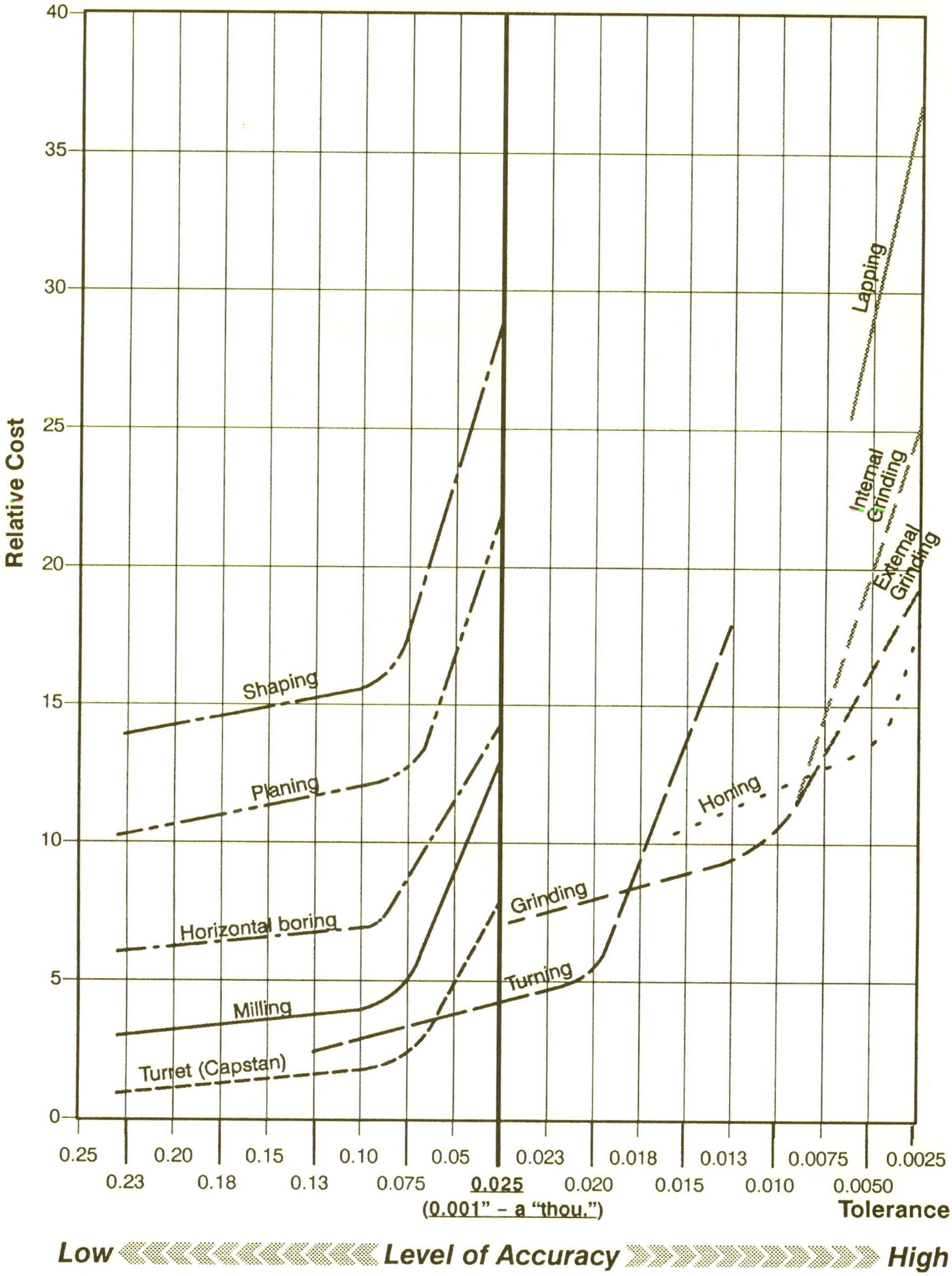


(from BS PD6470, 1981)

## 10.2 Design details for casting

### *10.2.1 Introduction*

Casting processes allow metal to be positioned according to the load to be carried by the components, thus making for more uniform stress distribution and good material utilization. The compressive strengths of casting alloys are considerably higher than their tensile strengths since, when in tension, discontinuities in the grain structure, such as graphite flakes in grey cast iron, become stress-raising features.

Castings should therefore be designed keeping the material in compression where possible, as shown in the figure. When bending is unavoidable the tension side can be reinforced using additional metal. Ribs on the tension side may help: their thickness should be 0.8–0.9 times the wall thickness and their depth 1.1–1.2 times the wall thickness.

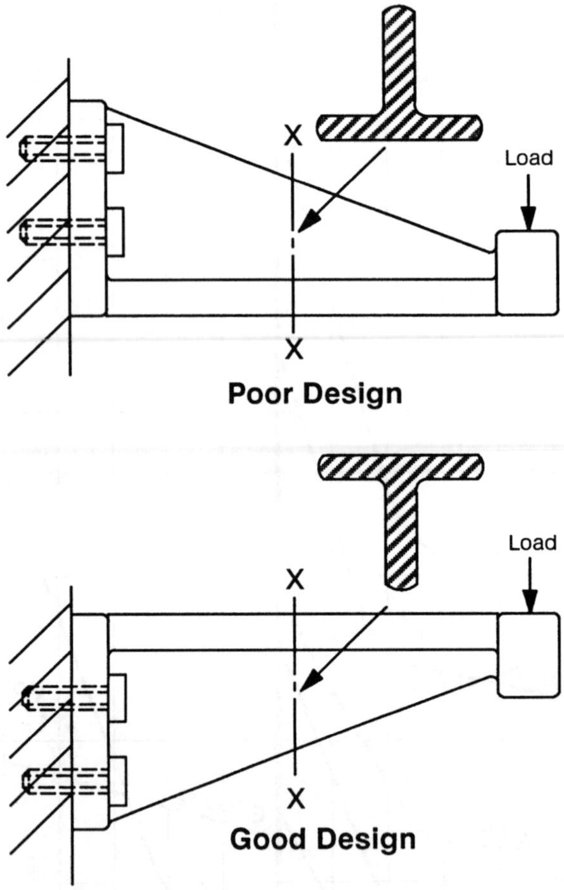


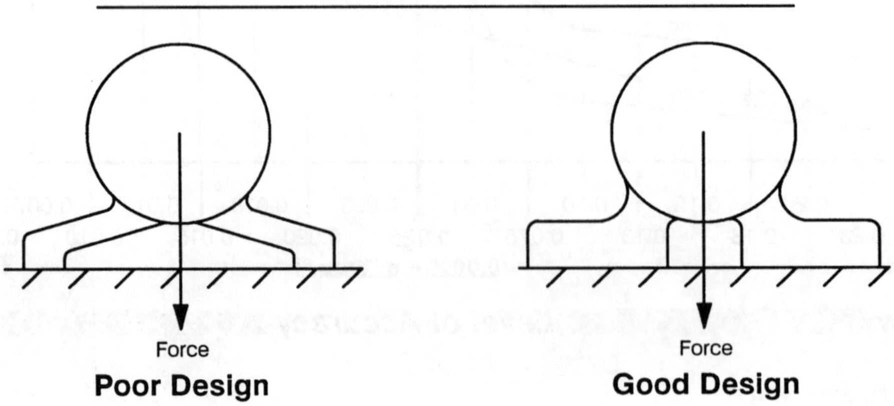

Different casting sections are preferred for different load conditions. For transverse bending, asymmetric 'I' sections give best use of material; for torsion, closed circular sections are best. For combined bending and twisting asymmetric closed sections are the optimum compromise.

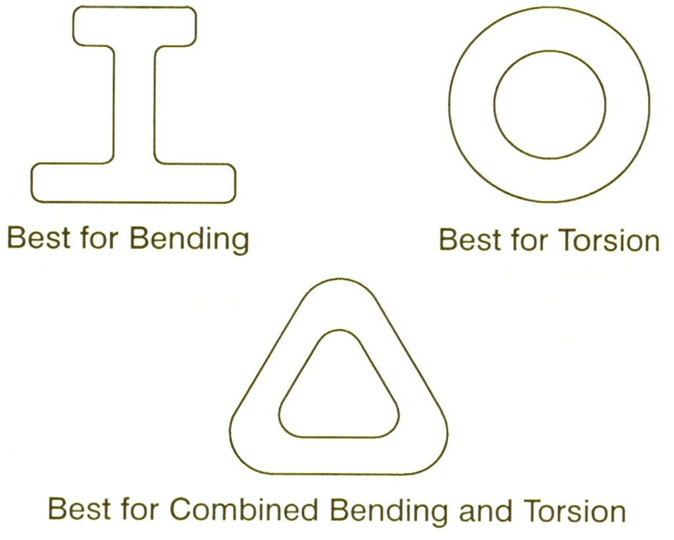


(All three sections have the same cross-sectional area and same height)

### 10.2.2 General design rules

*1 Aim for uniform grain size and structure without residual stress*

This is done by designing for a uniform rate of cooling – keep the sections as uniform as possible and blend the changes in section lengthwise; do not allow 'step' changes in features.

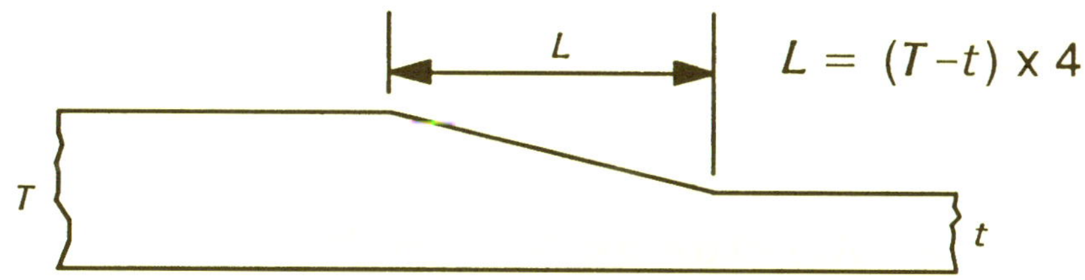


Also reduce heavy sections at junctions of sections by keeping fillet radii small (abount 0.5$t$) and by staggering them if possible.

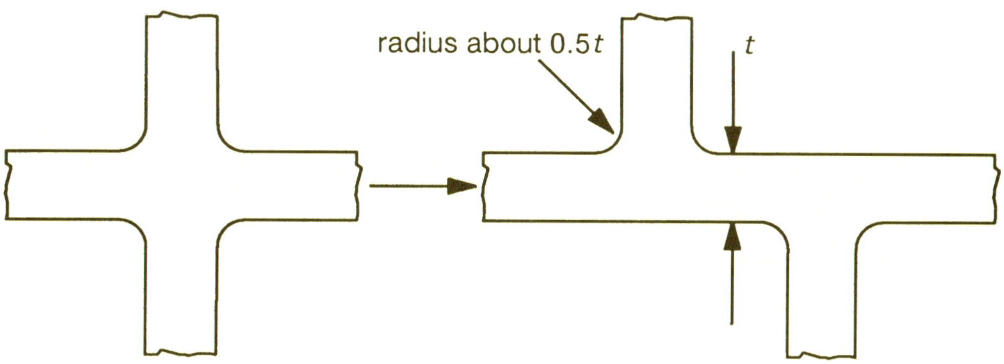

*2 Avoid metallurgical defects*

(a) Avoid sharp corners with their associated plane of weakness where columnar crystals inhibit each others' development (outside corner) or grow away from each other (inside corner), as shown in the following figure. The extent of formation of the columnar crystals is primarily a function of the type of metal used, the solidification rate and the shape of the casting.

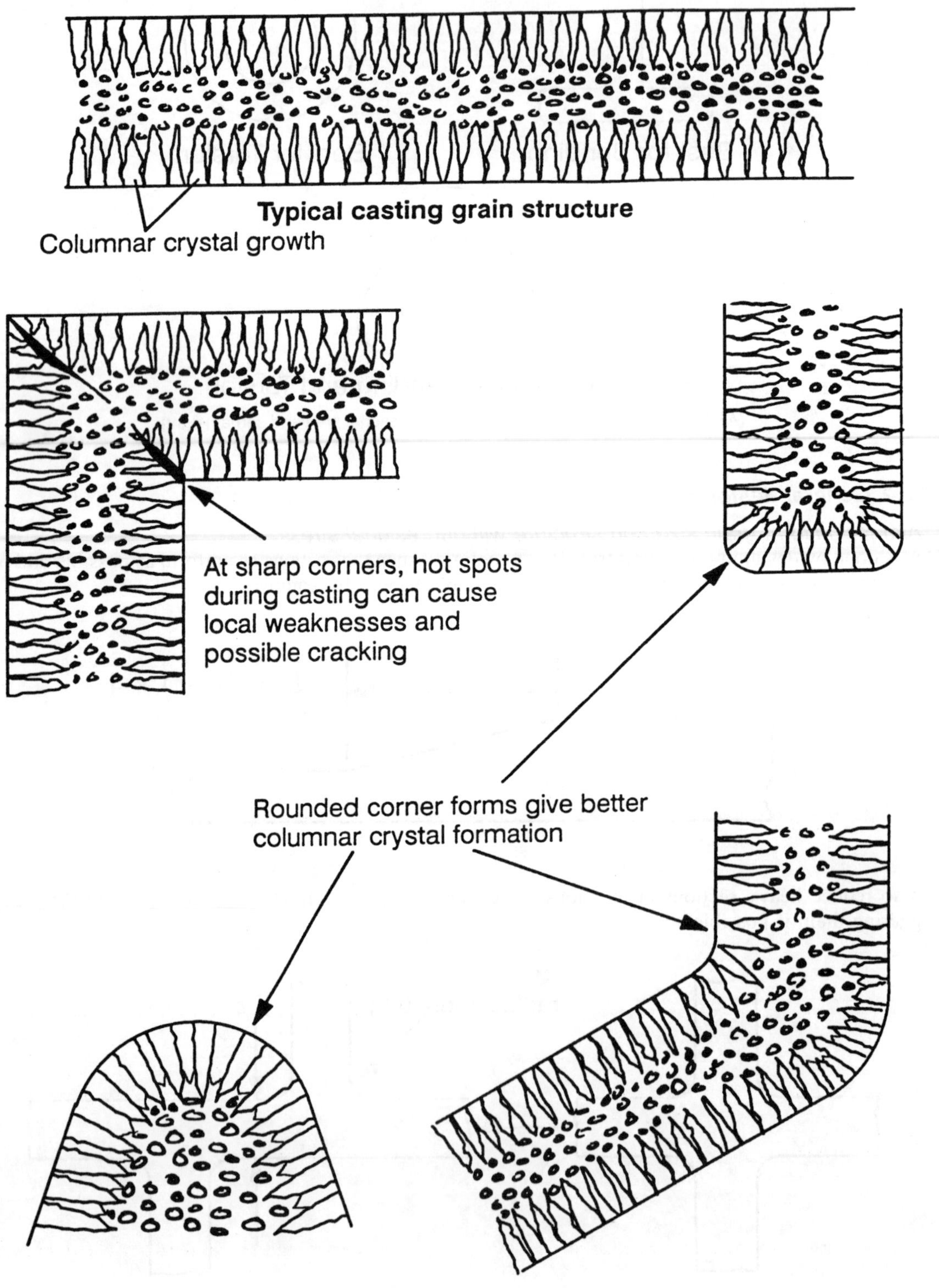

(b) Reduce residual stresses and the risk of hot tears by designing for flexibility. Try to design to minimize cooling stresses as shown in the next figure.

Rim susceptible to hot tearing or residual compressive stresses. The spokes may be susceptible to tearing or residual tensile stresses.

Spokes susceptible to tearing or residual compressive stresses; the rim may be susceptible to residual tensile stresses and flattening.

This wheel has a balanced section thickness which gives uniform cooling.

*3 Design for ease of moulding*

(a) Try to reduce the number of external and internal cores.

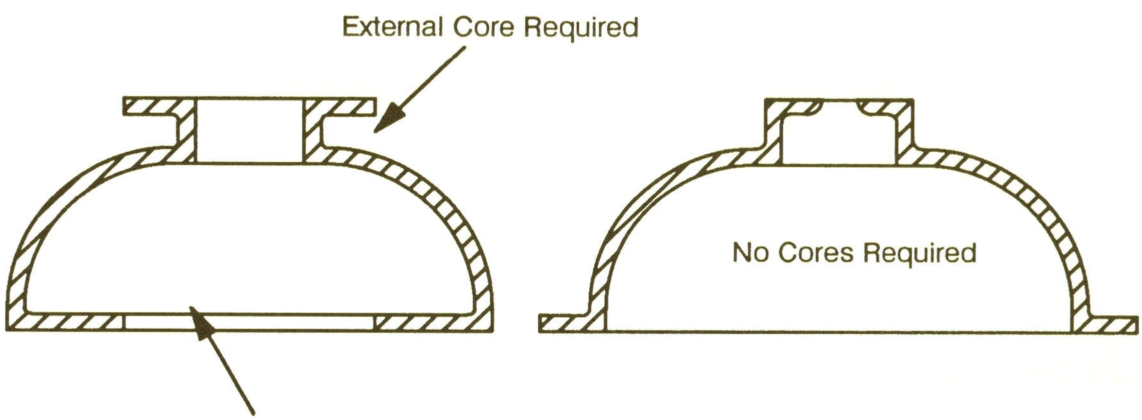


(b) Avoid large areas of thin sectioned castings where surface to volume ratio is high and premature solidification is likely.

*4 Design for ease of machining*

(a) Minimize the areas to be machined.
(b) Arrange for all surfaces to be at the same height.

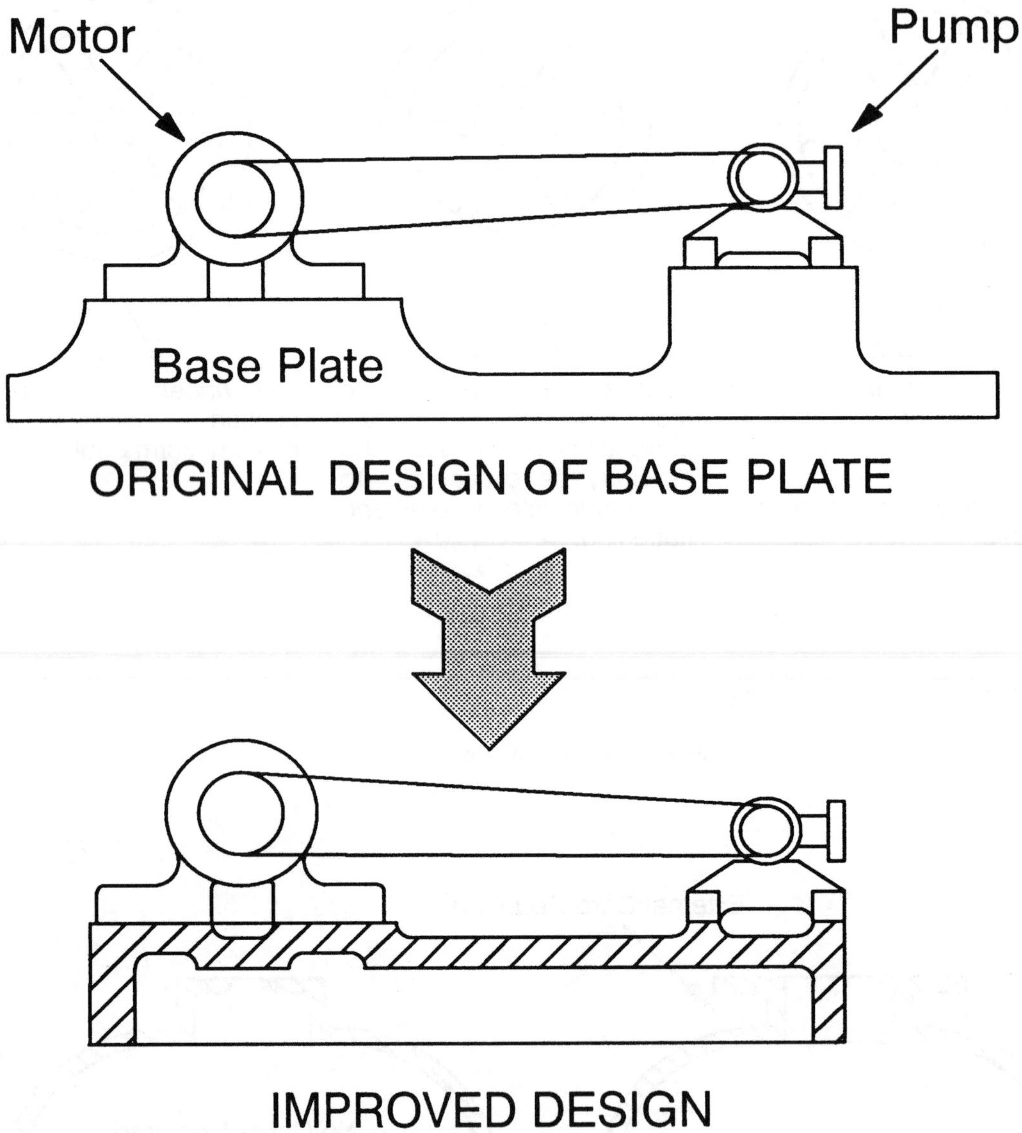


## 10.3 Design details for machining

Careful attention to detail can reduce the cost a component contributes to an assembly by reducing costs in a number of areas; some of these are described below.

### *10.3.1 Cost of material*

The cost of material can be reduced by designing to utilize standard bar sizes held in stock by the company. Choosing a free-machining grade of the required specification may increase the material costs slightly but can show a substantial reduction in machining costs later on.

### 10.3.2 Cost of machining

The cost of machining can be reduced by designing with the following aims in mind:

(a) Design so that the component can be machined on as few machines as possible with a minimum of set-ups.
(b) Minimize the area to be machined by designing castings, forgings and fabrications in such a way that only those areas which need to be machined are left standing proud, as shown in the figure. This can be especially worthwhile when tight tolerances on dimensions and surface finish have to be achieved.

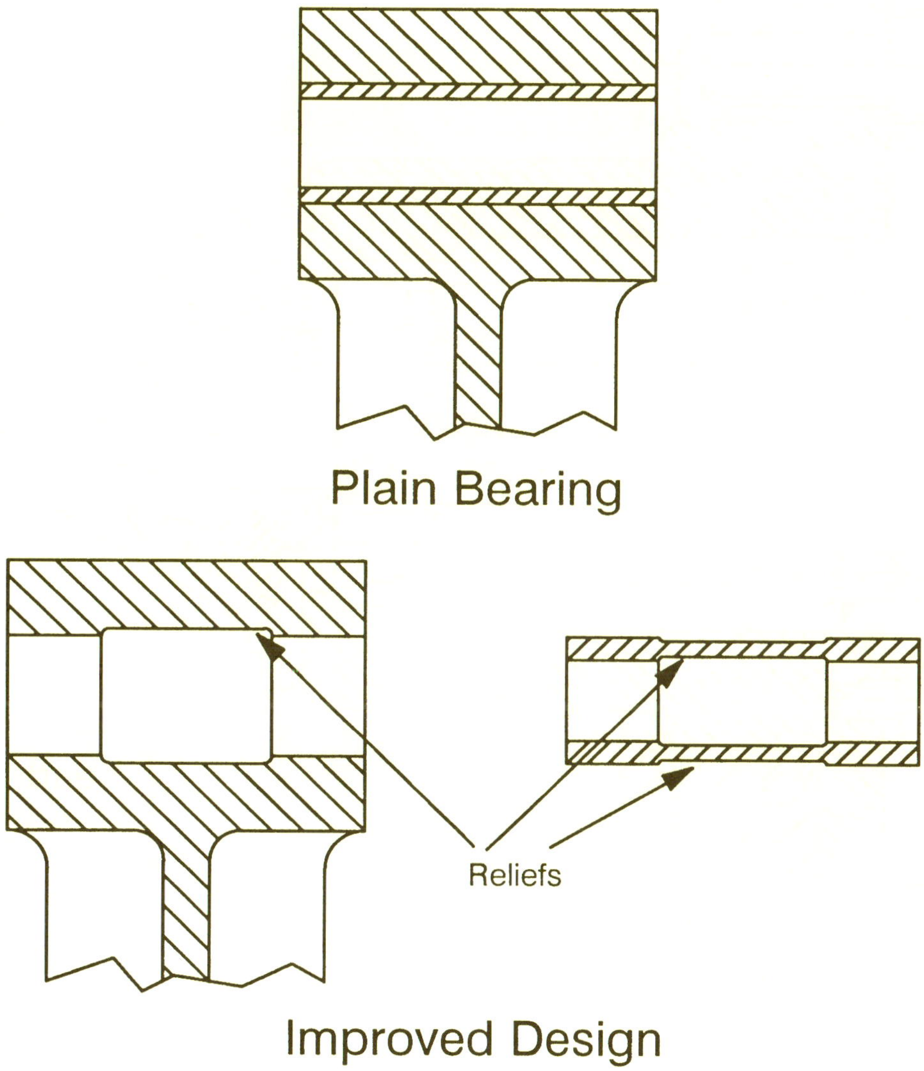


(c) Allow high metal removal rates by avoiding features which cannot be machined by standard coated carbide cutting tool. These can remove material 3–5 times faster than high-speed steel tools which are used for non-standard shapes.
(d) Avoid long slender sections, thin walls, deep bores and other features which give rise to undue deflections of the workpiece or the tool; the component may consequently have to be machined at reduced feed rates.

### 10.3.3 Cost of assembly

The cost of assembly can be reduced by selecting ball races and other pressed on/in parts so that they only have to be pressed along the shaft or into the bore in which they will ultimately fit. Adequate leads on the shafts and housings can also help to reduce assembly time.

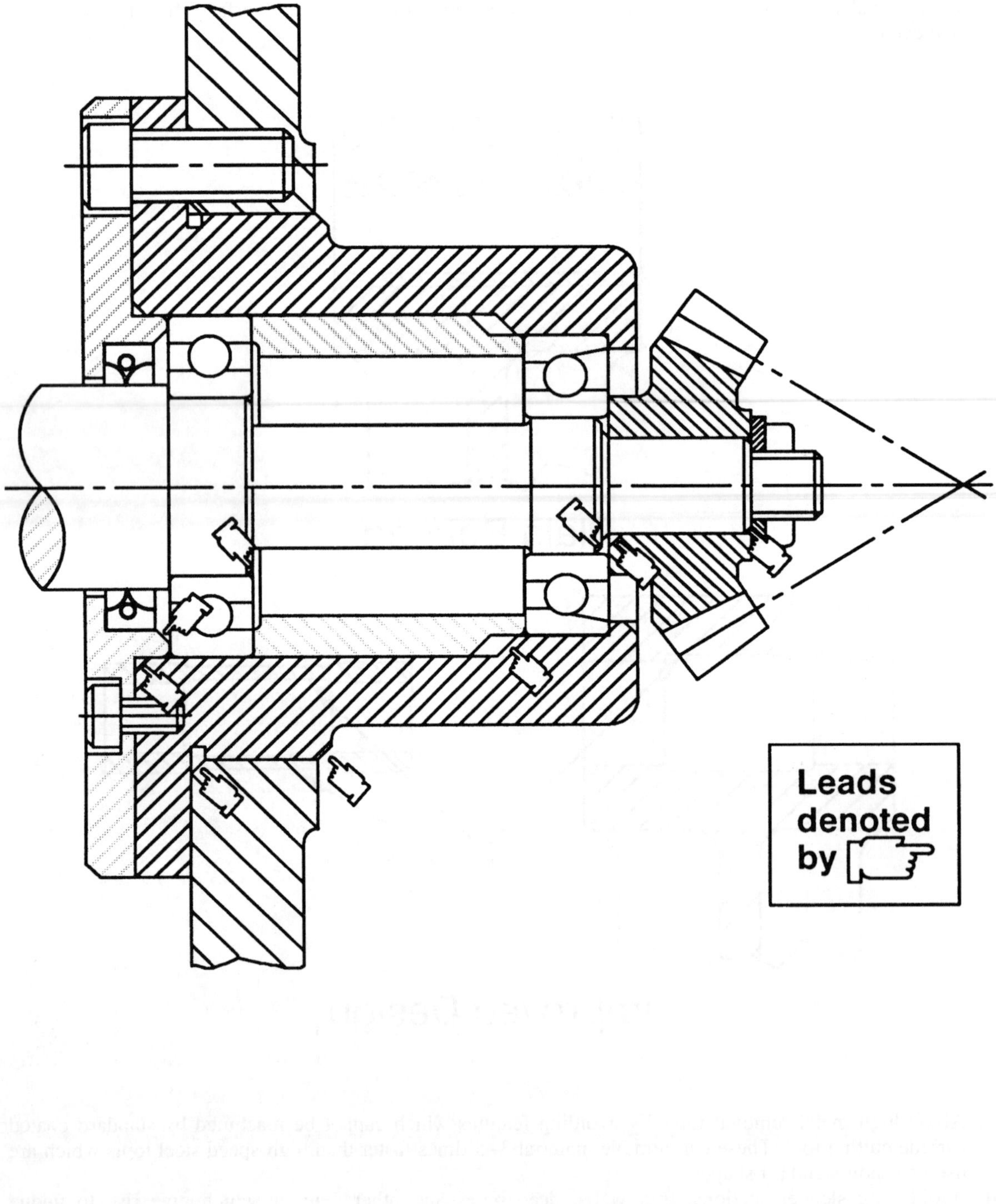

### *10.3.4 Particular details for machining*

(a) Slender cylindrical components (length: diameter ratio > 3) will have to be supported by a centre at the tailstock end when being turned. Thus internal features cannot be machined at the same set-up.
(b) Tools such as twist drills and slot drills are only effective when their length: diameter ratio is less than 5 or 6, so avoid features which are inaccessible to standard tools.

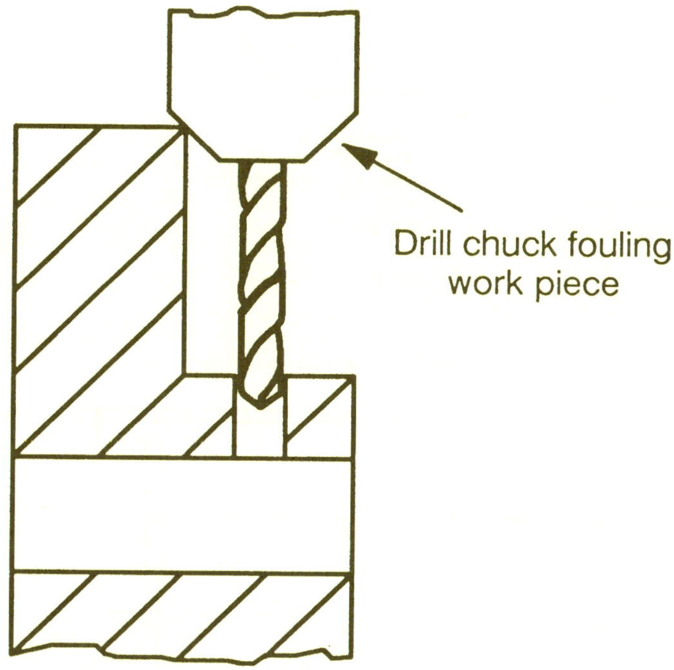


(c) Grinding wheels quickly develop corner radii and therefore cannot grind sharp internal corners such as when a diameter runs up to a shoulder on a shaft. Undercutting is a solution which may introduce a stress-raising feature; it is better therefore to chamfer or counter-bore the mating part.

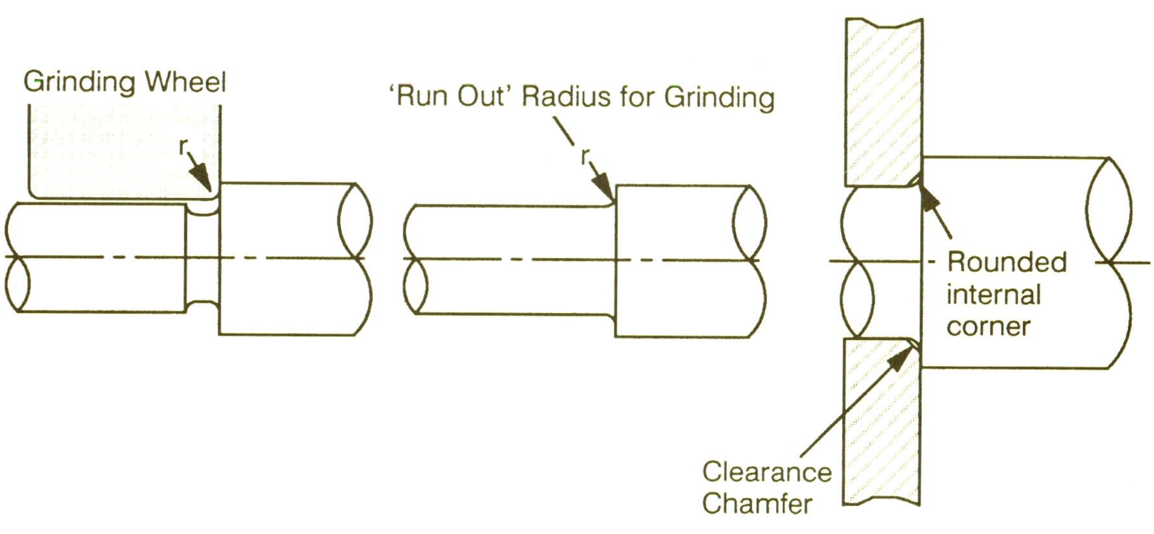

(d) When screw threads run up to a shoulder, the last two or three pitches will not be fully formed. Undercutting is a solution but, as previously, counter-boring the mating part is usually the better course of action. The following figure shows both solutions.

Profile of undercuts

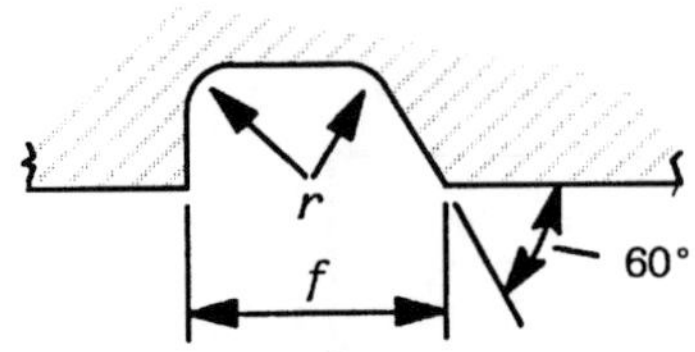


$p$ = thread pitch
$d$ = thread diameter
$g$ = thread core diameter

$r = 0.5 \times p$

$f = 3 \times p$

Example for an M16 thread

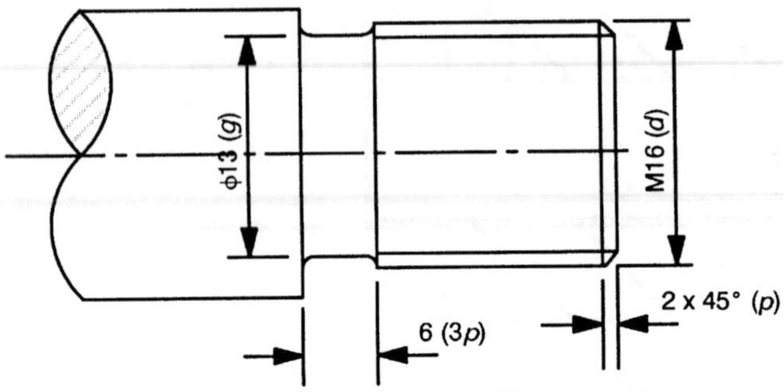


*Undercut and chamfer with external screw thread adjacent to a shoulder*

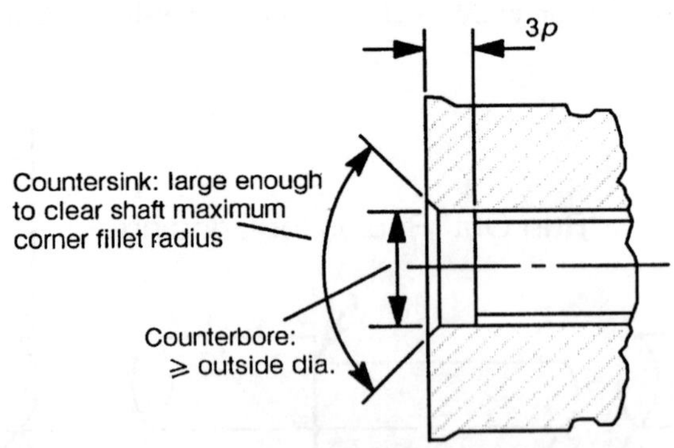


*Alternative to an undercut on the external screwthread*

(e) Avoid drilling blind holes for tapping and reaming where possible. They cannot be fully tapped or reamed, they are time-consuming and the risk of tool breakage is increased.

(f) Drills should enter and break through normal to the surfaces involved to avoid the risk of tool breakage.

(g) Standard slot drills and end mills do not have corner radii so the bottom internal corners of workpieces should be sharp; other internal corners should have as large radii as possible to allow the maximum diameter of cutter to be used. This is illustrated in the next figure.

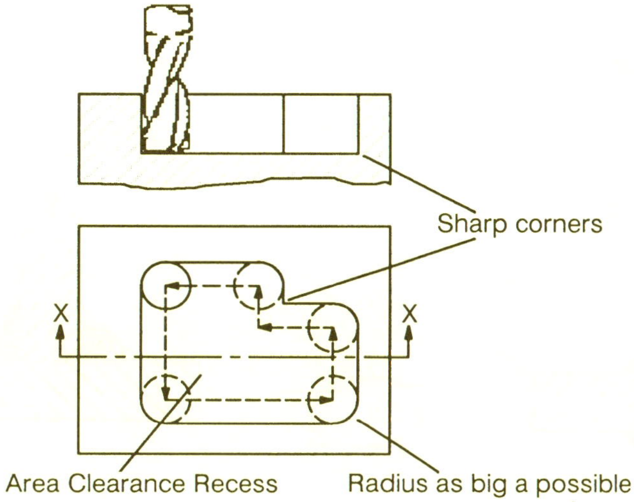


## 10.4 Design details for welding

The cost of welding can be reduced and the effectiveness of weldments increased if the simple rules outlined below are understood and applied wherever possible.

### *10.4.1 Minimum weld preparations*

Chamfering of edges, whether by grinding, machining or flame-cutting, is time-consuming and adds to the overall cost of weldments. The examples illustrated below show how weld preparations can often be avoided. Note, however, that not only does the appearance of the finished job change but also the loading on the weld.

(a)

(b)

(c)

If access is possible from both sides then weld preparation on both sides (right-hand diagram) reduces the amount of metal to be removed by 50%; consequently the weld requires only 50% of the weld material to be deposited compared with single-sided preparation. This is an important consideration for plate thicknesses above 5 mm.

### 10.4.2 Design for location

(a)

Weld carries load

Weld for location only

(b)

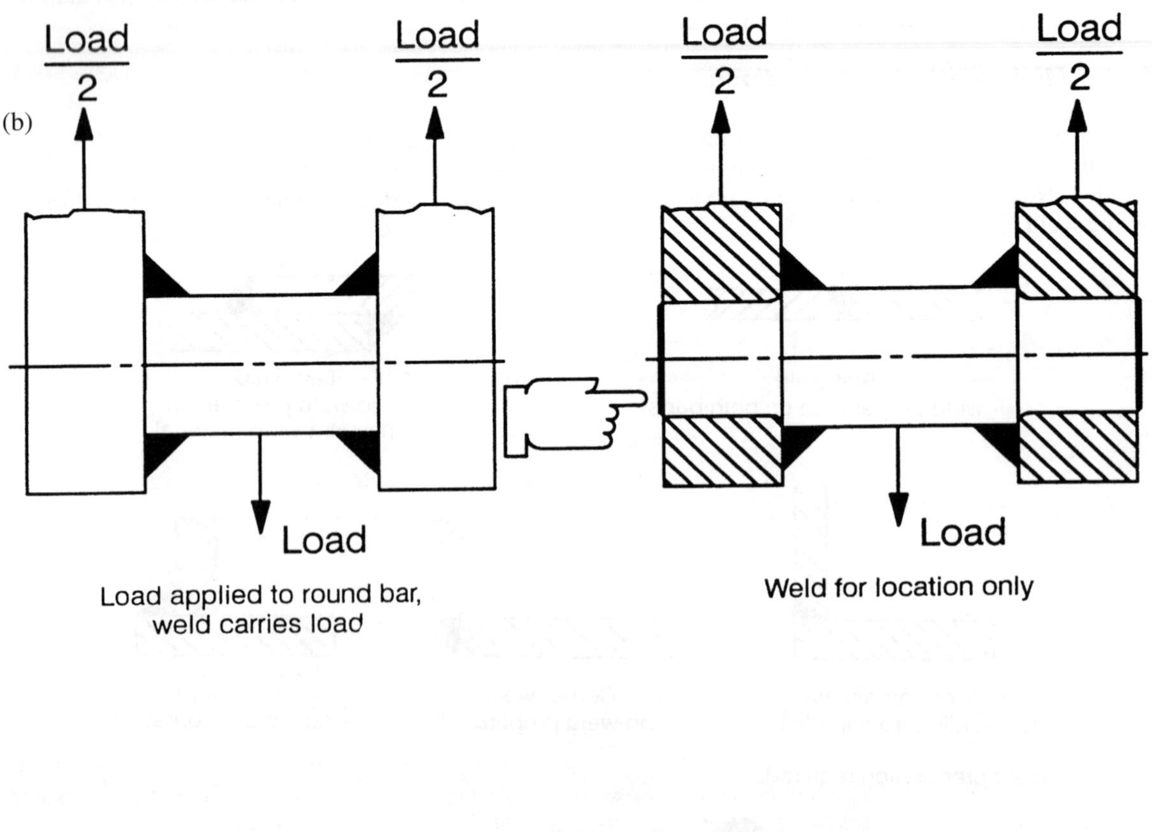


Load applied to round bar, weld carries load

Weld for location only

### *10.4.3 Positioning load-carrying welds for maximum effectiveness*

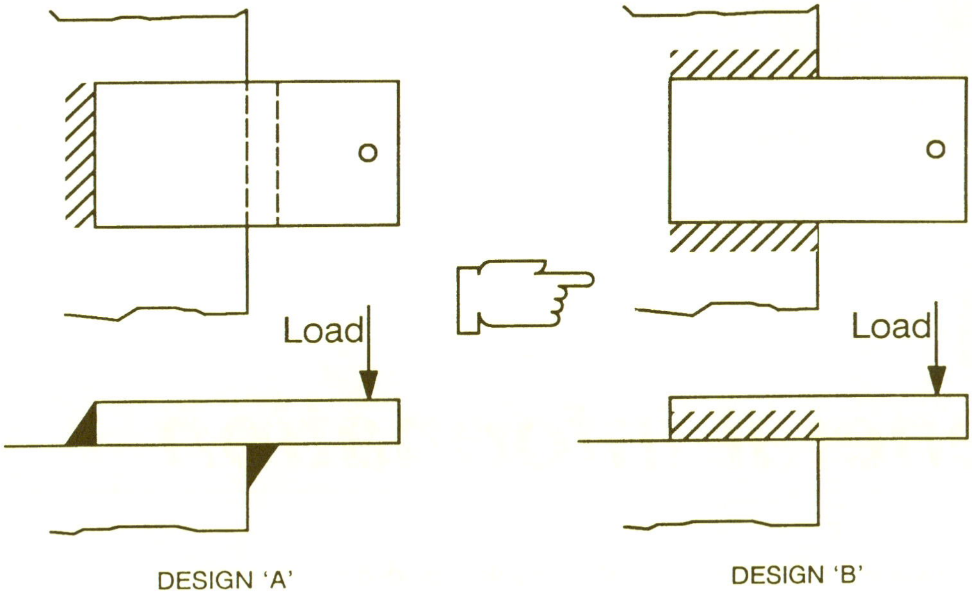


Note: 1) For design 'A' the right hand weld is for location only.

2) For equal strength Design 'B' requires three times as much weld as Design 'A'.

### *10.4.4 Allow for weld shrinkages*

The weld metal deposited shrinks on solidification and cooling to ambient temperature. This can lead to distortion, residual stresses and cracking if not taken into consideration.

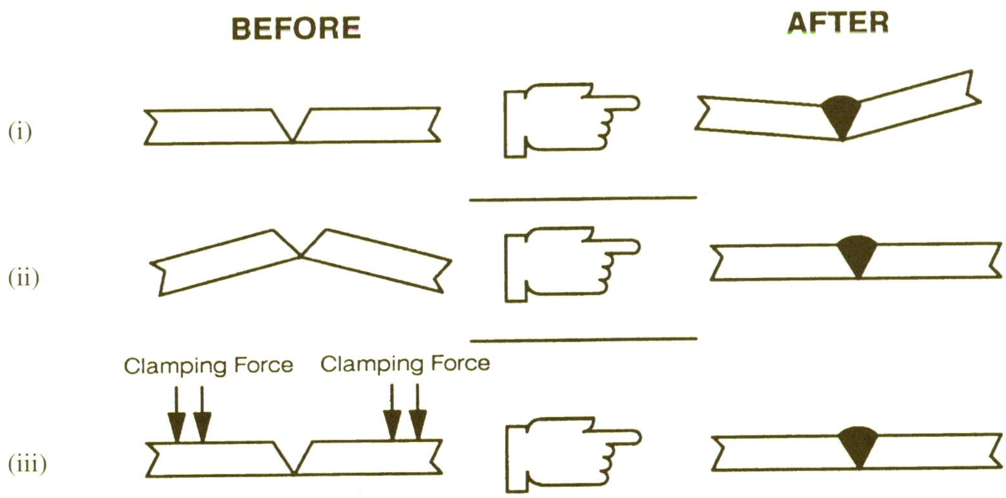


The welding set-up illustrated in (ii) above allows shrinkage to pull the job straight. The flexibility of the set-up eliminates the risk of cracks arising from hot tears and minimizes residual stresses.

If the parts to be joined are rigidly clamped during welding, distortion is eliminated but there is a risk of weld cracking and a certainty of residual stresses which will reduce the load-carrying capability of the finished job. Post-weld stress-relieving heat treatment is sometimes necessary.

# 11
# *General information*

## 11.1 Engineering mathematical formulae and definitions

$\pi = 3.141\,592\,6536$

### *11.1.1 Lengths, areas and volumes*

*11.1.1.1 Planes*

Square

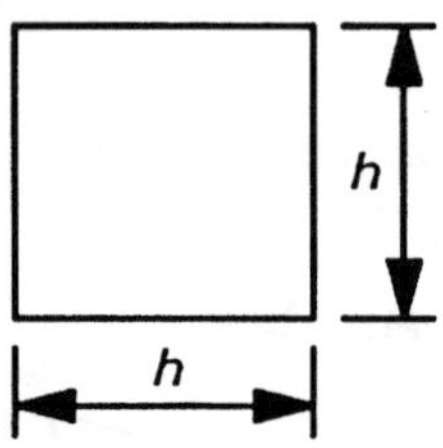


area $= h^2$

length of diagonal $= \sqrt{2}\,h$

Rectangle

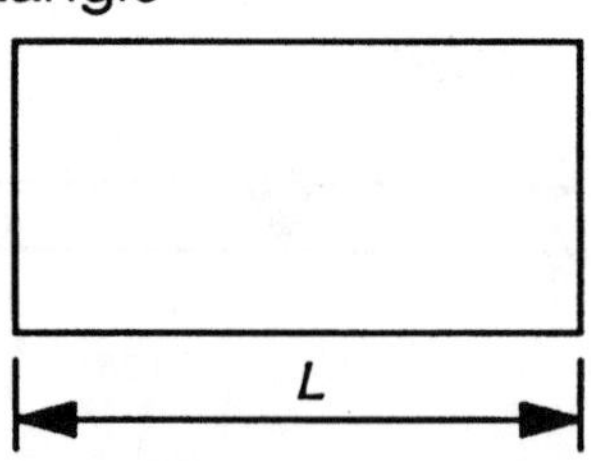


area $= L \times h$

length of diagonal $= \sqrt{(L^2 + h^2)}$

Parallelogram

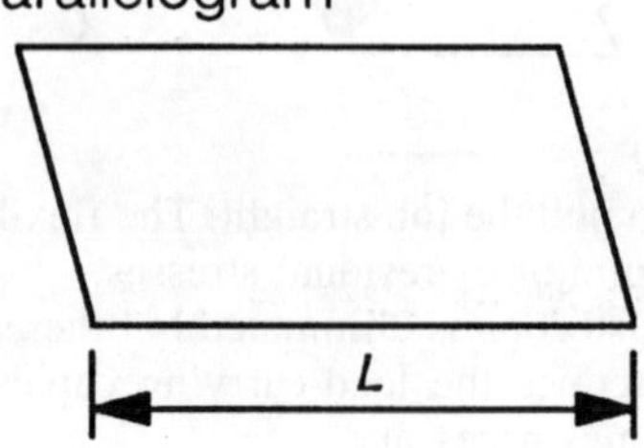


area $= Lh$

Trapezium

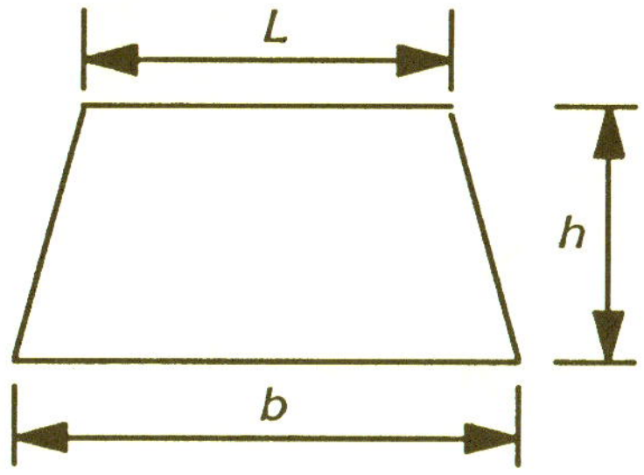


$$\text{area} = \frac{1}{2}(L + b)h$$

Triangles

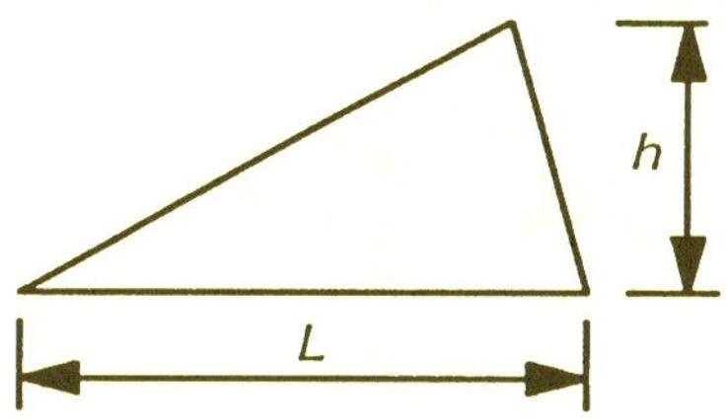


$$\text{area} = \frac{1}{2}Lh$$

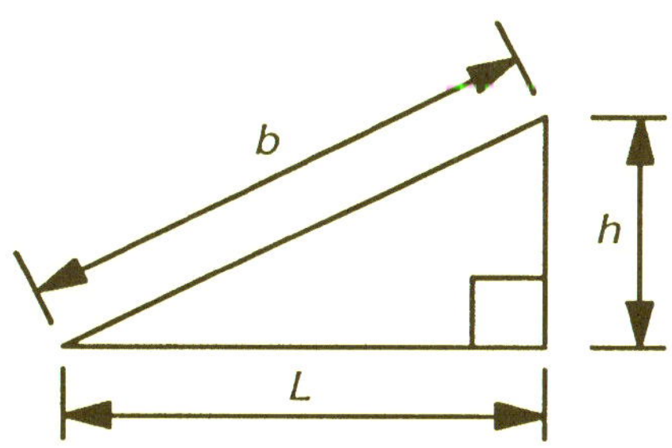


$$b = \sqrt{L^2 + h^2}$$

Ellipse

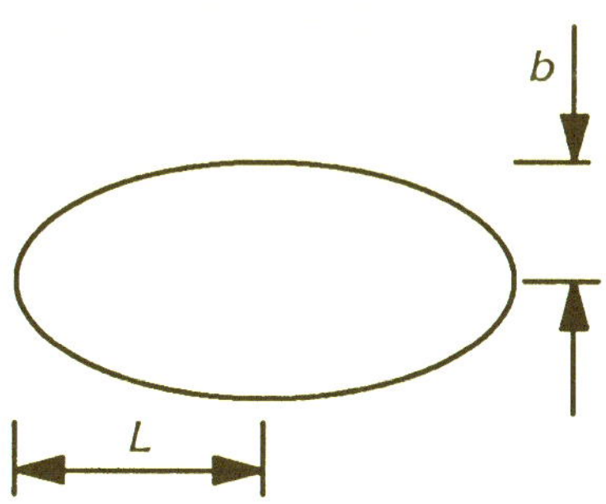


$$\text{area} = \pi L b$$

$$\text{perimeter} = \pi(L + b)$$

*11.1.1.2 Circle definitions and properties*

$$\text{area} = \pi r^2 = \frac{\pi d^2}{4}$$

$$\text{circumference} = 2\pi r = \pi d$$

The *arc* of a circle is a part of the circumference.

A *tangent* to a circle is a straight line which meets the circumference at one point only. A line drawn at right angles to the tangent at the point where it meets the circumference will pass through the centre of the circle.

A Tangent

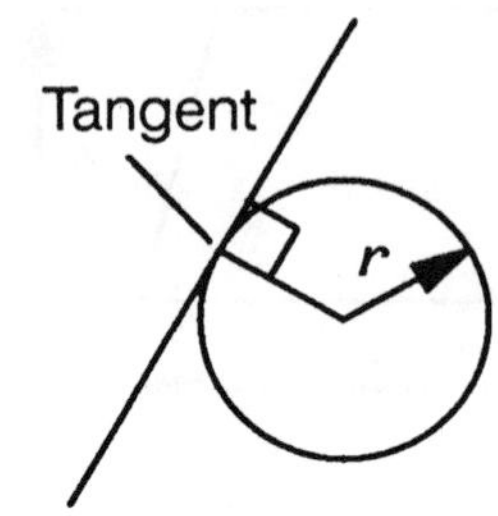


A *sector* of a circle is the area bounded by a circle arc and two radii.

Circle Sector

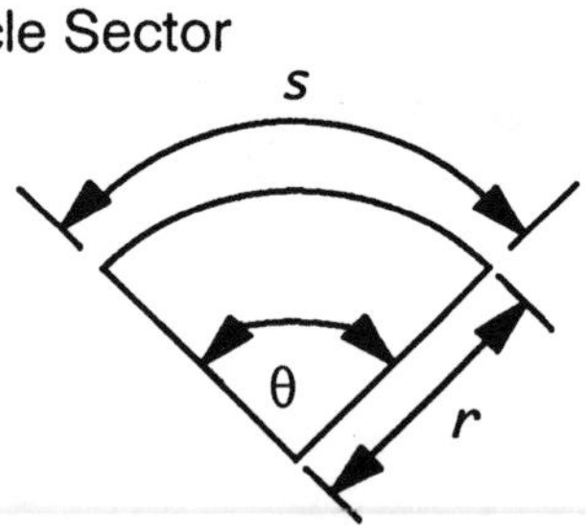


$$\text{area} = \frac{1}{2}r^2\theta$$

$$\text{arc length } s = r\theta$$

N.B. – $\theta$ is in radians

A *chord* is a straight line joining two points on the circumference of a circle.

Circle Chords

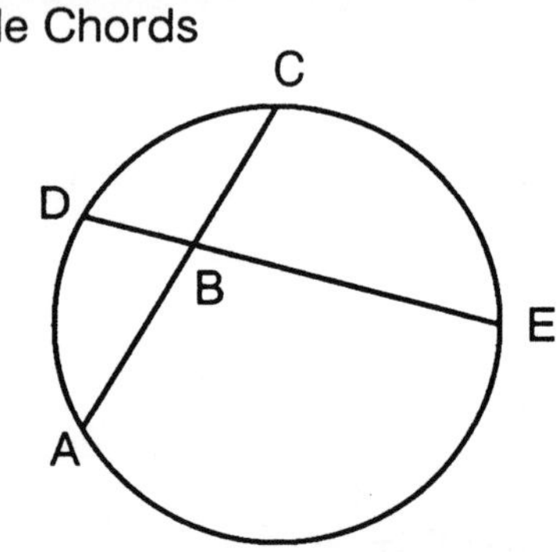


$$AB \times BC = DB \times BE$$

A *segment* is the area bounded by a chord and an arc. Angles in the same segment of a circle are equal.

Circle Segments

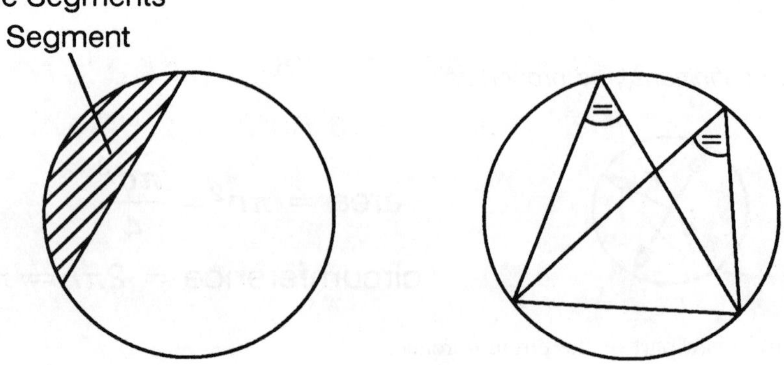

*11.1.1.3 Solids*

**Cuboid**

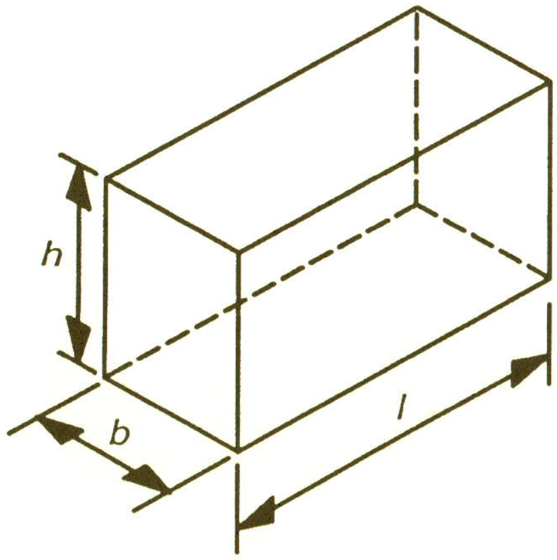


Volume $= bhl$

Total Surface Area $= 2(bh + hl + lb)$

**Cylinder**

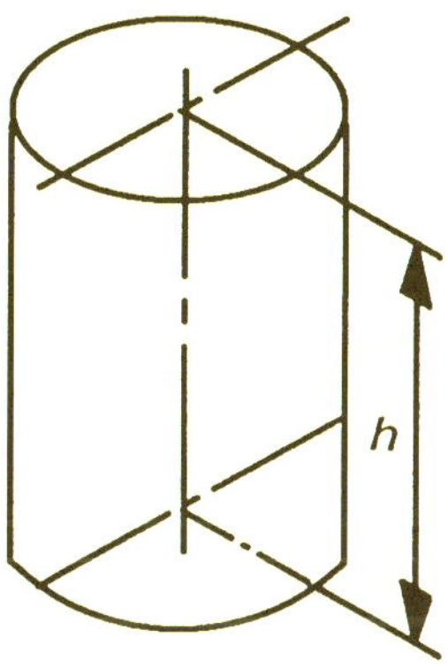


Volume $= \pi r^2 h$

Total Surface Area $= 2\pi r(r + h)$

**Cone**

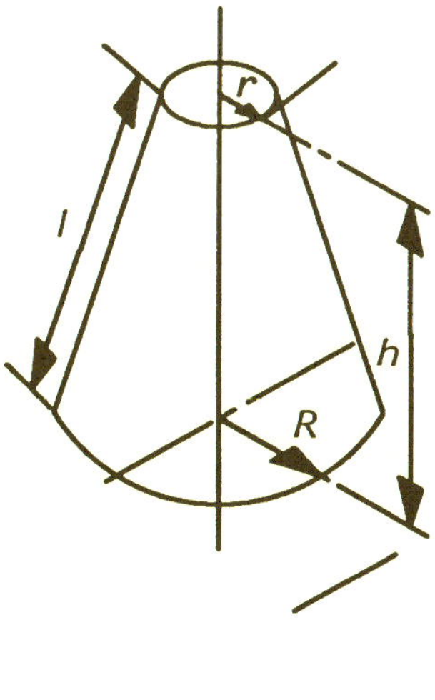


Volume $= \frac{1}{3}\pi h(R^2 + Rr + r^2)$

Total Surface Area $= \pi l(R + r) + \pi(R^2 + r^2)$

## Sphere

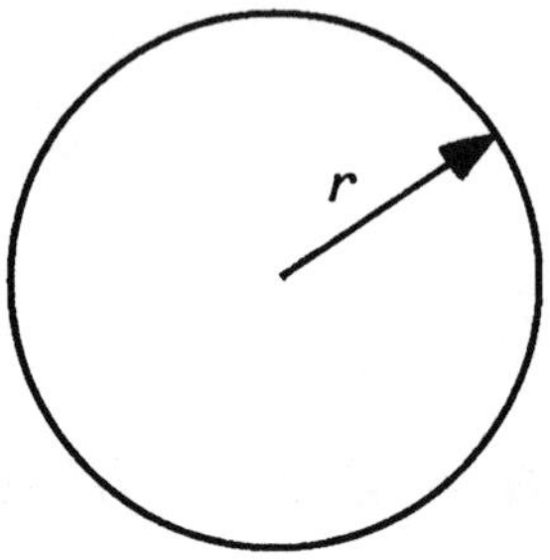


$$\text{Volume} = \frac{4}{3}\pi r^3$$

$$\text{Total Surface Area} = 4\pi r^2$$

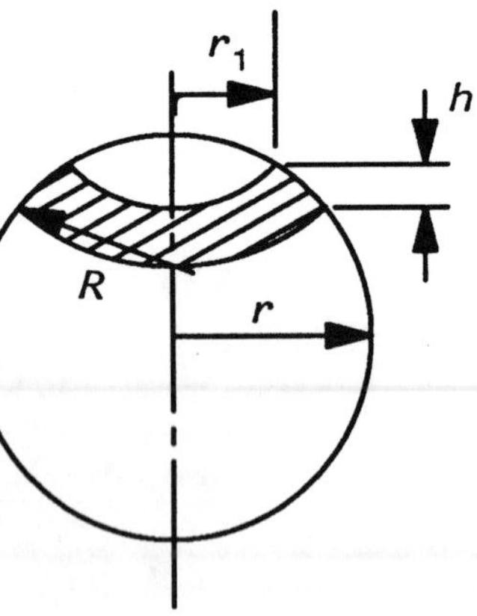


$$\text{Volume} = \left(\frac{\pi h}{6}\right)(h^2 + 3R^2 + 3r_1^2)$$

$$\text{Total Surface Area} = 2\pi rh + \pi(R^2 + r_1^2)$$

## Pyramid

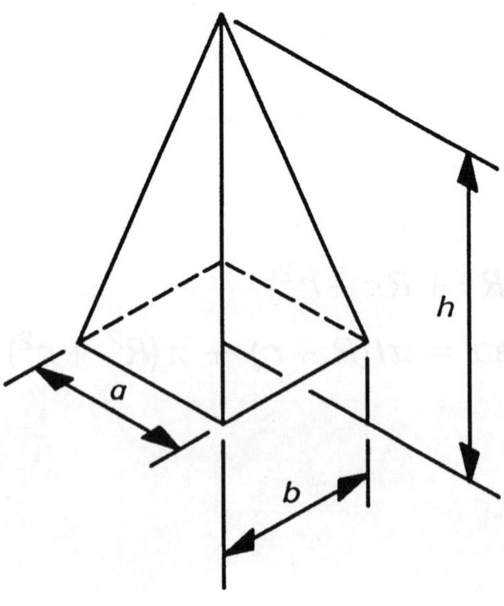


$$\text{Volume} = \frac{1}{3}abh$$

### 11.1.2 Commonly used trigonometric formulae

*Right-angled triangle*

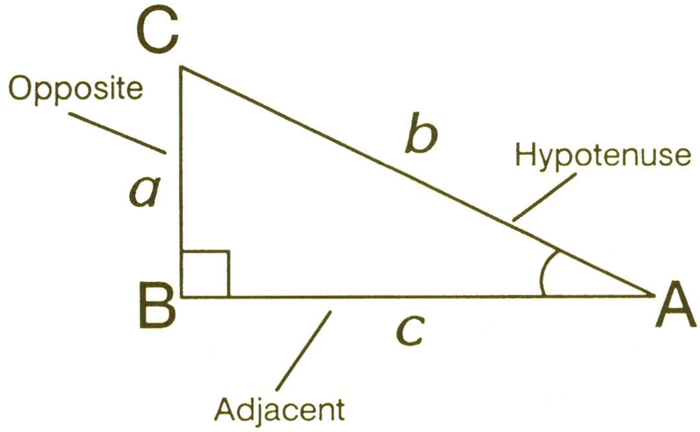


*Basic trigonometric definitions*

$$\sin A = \frac{\text{opposite}}{\text{hypotenuse}} = \frac{a}{b}$$

$$\cos A = \frac{\text{adjacent}}{\text{hypotenuse}} = \frac{c}{b}$$

$$\tan A = \frac{\text{opposite}}{\text{adjacent}} = \frac{a}{c}$$

$$\operatorname{cosec} A = \frac{\text{hypotenuse}}{\text{opposite}} = \frac{b}{a} = \frac{1}{\sin A}$$

$$\sec A = \frac{\text{hypotenuse}}{\text{adjacent}} = \frac{b}{c} = \frac{1}{\cos A}$$

$$\cot A = \frac{\text{adjacent}}{\text{opposite}} = \frac{c}{a} = \frac{1}{\tan A}$$

*Common identities*

$$\sin^2 A + \cos^2 A = 1$$

$$1 + \tan^2 A = \sec^2 A$$

$$1 + \cot^2 A = \operatorname{cosec}^2 A$$

$$\sin(-A) = -\sin A$$

$$\cos(-A) = \cos A$$

$$\tan(-A) = -\tan A$$

*Compound and double angle formulae*

$$\sin(A + B) = \sin A \cos B + \cos A \sin B$$

$$\sin(A - B) = \sin A \cos B - \cos A \sin B$$

$$\cos(A + B) = \cos A \cos B - \sin A \sin B$$

$$\sin 2A = 2 \sin A \cos 2A$$

$$\cos(A - B) = \cos A \cos B + \sin A \sin B$$

$$\tan(A + B) = \frac{\tan A + \tan B}{1 - \tan A \tan B}$$

$$\tan(A - B) = \frac{\tan A - \tan B}{1 + \tan A \tan B}$$

$$\tan 2A = \frac{2 \tan A}{1 - \tan^2 A}$$

$$\cos 2A = \cos^2 A - \sin^2 A = 2\cos^2 A - 1 = -2\sin^2 A$$

*Product formulae*

$$\sin A \cos B = \frac{1}{2}[\sin(A+B) + \sin(A-B)] \qquad \cos A \sin B = \frac{1}{2}[\sin(A+B) - \sin(A-B)]$$

$$\cos A \cos B = \frac{1}{2}[\cos(A+B) + \cos(A-B)] \qquad \sin A \sin B = -\frac{1}{2}[\cos(A+B) + \cos(A-B)]$$

ACTS!

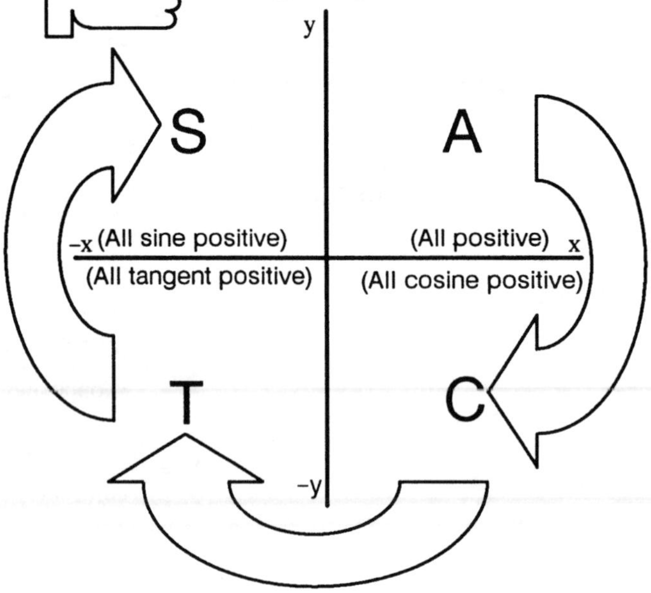


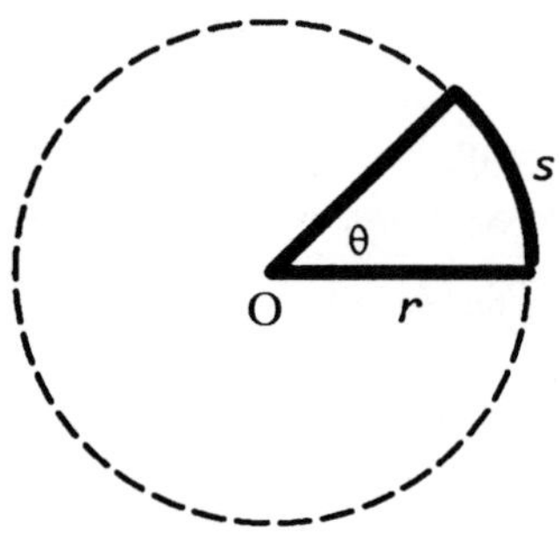


In radians: $\theta = \dfrac{s}{r}$

Thus: $s = \theta r$

$s = 2\pi r$

For a circle of $r$ = units radius:

$s = \theta = 2\pi$

Therefore:

$360° = 2\pi$ radians, $180° = \pi$ radians, $90° = \pi/2$ radians

Thus: 1 radian = 57° 17' 44.8"

*11.1.2.2 Any triangle*

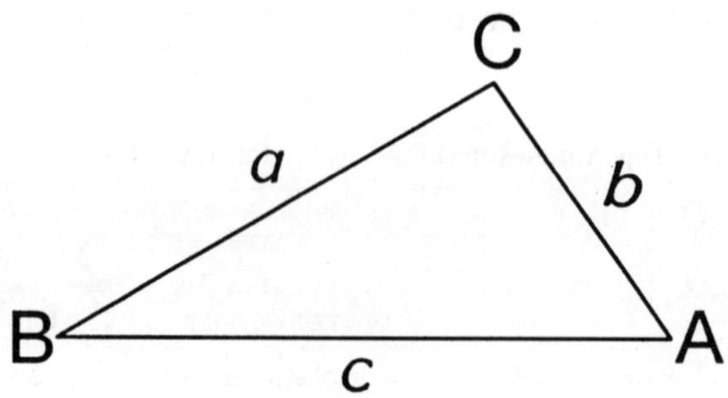

*Basic formulae*

*Sine rule* $$\frac{a}{\sin A} = \frac{b}{\sin B} = \frac{c}{\sin C}$$

*Cosine rule* $$a^2 = b^2 + c^2 - 2bc \cos A$$
$$b^2 = a^2 + c^2 - 2ac \cos B$$
$$c^2 = a^2 + b^2 - 2ab \cos C$$

*Area* $$\text{area} = \frac{1}{2} ab \sin C = \frac{1}{2} ac \sin B = \frac{1}{2} bc \sin A$$
$$\text{area} = \sqrt{s(s-a)(s-b)(s-c)}$$

where $s$ is the semi-perimeter

$$s = \frac{(a+b+c)}{2}$$

### *11.1.3 The Greek Alphabet – with common engineering usages*

| Description | *Letter* Capital | Lower case | Engineering and scientific uses |
|---|---|---|---|
| alpha | Α | α | various coefficients, angles, angular acceleration |
| beta | Β | β | various coefficients, angles |
| gamma | Γ | γ | kinematic viscosity, shear strain, surface tension |
| delta | Δ | δ | damping coefficient, differences, derivatives |
| epsilon | Ε | ϵ | linear strain |
| zeta | Ζ | ζ | |
| eta | Η | η | efficiency, dynamic viscosity |
| theta | Θ | θ | angles, temperature, volume strain |
| iota | Ι | ι | |
| kappa | Κ | κ | compressibility |
| lambda | Λ | λ | wavelength, thermal conductivity |
| mu | Μ | μ | coefficient of friction, Poisson's ration |
| nu | Ν | ν | dynamic viscosity |
| xi | Ξ | ξ | |
| omicron | Ο | ο | |
| pi | Π | π | mathematical constant = 3.1415926536 |
| rho | Ρ | ρ | density |
| sigma | Σ | σ | normal stress, standard deviation, sum of |
| tau | Τ | τ | shear stress |
| upsilon | Υ | υ | |
| phi | Φ | φ | angles, heat flow rate, potential energy, diameter |
| chi | Χ | χ | |
| psi | Ψ | ψ | helix angle |
| omega | Ω | ω | electrical resistance, angular velocity |

## 11.2 Units and common constants

### 11.2.1 Decimal multiples of units

| *Multiple* | *Name* | *Symbol* |
|---|---|---|
| $10^{-18}$ | atto | a |
| $10^{-15}$ | femto | f |
| $10^{-12}$ | pico | p |
| $10^{-9}$ | nano | n |
| $10^{-6}$ | micro | μ |
| $10^{-3}$ | milli | m |
| $10^{-2}$ | centi | c |
| $10^{-1}$ | deci | d |
| 10 | deca | da |
| $10^{2}$ | hecto | h |
| $10^{3}$ | kilo | k |
| $10^{6}$ | mega | M |
| $10^{9}$ | giga | G |
| $10^{12}$ | tera | T |
| $10^{15}$ | peta | P |
| $10^{18}$ | exa | E |

*Note*: Compound unit multiples on one unit must not be used, e.g. Mg not kkg.

### 11.2.2 Basic units

All the units used in this book relate to the International System of Units (SI). Conversions are given later in this section from SI to Imperial or Foot-Pound-Second (FPS) units.

The main fundamental physical properties on which all other units are based are shown in the table.

| *Physical property* | *Unit* | *Abbreviation* |
|---|---|---|
| length | metre | m |
| mass | kilogram | kg |
| time | second | s |
| plane angle | radian | rad |
| amount of substance | mole | mol |
| electric current | ampere | A |
| luminous intensity | candela | cd |
| solid angle | steradian | sr |
| thermodynamic temperature | kelvin | K |

### 11.2.3 Derived units

By combinations of multiplying and/or dividing the units above, derived SI units for specific physical properties can be obtained.

Where the units of such a property are given a special name, the derivation is given to clarify how they were developed. Many special names are given to units in recognition of the contributions made to engineering and science by famous engineers, mathematicians and scientists. These units are also especially convenient where the derivation is quite complicated and the units themselves not easy to remember.

| *Physical property* | *Unit symbol* | *Derivation* |
|---|---|---|
| acceleration | $m/s^2$ | – |
| angular velocity | rad/s | – |
| area | $m^2$ | – |
| current density | $A/m^2$ | – |
| density | $kg/m^3$ | – |
| dynamic viscosity | Pa s | – |
| electric capacitance | F (farad) | $(A^2 s^4)/(kg\, m^2)$ |
| electric charge | C (coulomb) | A s |
| electric charge density | $C/m^3$ | – |
| electric conductance | S (siemen) | $(A^2 s^3)/(kg\, m^2)$ |
| electric field strength | V/m | – |
| electric potential difference | V (volt) | $(kg\, m^2)/(A\, s^3)$ |
| electrical resistance | Ω (ohm) | $(kg\, m^2)/(A^2 s^3)$ |
| energy | J (joule) | $(kg\, m^2)/s^2$ |
| energy density | $J/m^3$ | – |
| force | N (newton) | $(kg\, m)/s^2$ |
| frequency | Hz (hertz) | 1/s |
| heat capacity | J/K | – |
| heat flux density | $W/m^2$ | – |
| illuminance | lx (lux) | $(cd\, sr)/m^2$ |
| inductance | H (henry) | $(kg\, m^2)/(A^2 s^2)$ |
| luminance | $cd/m^2$ | – |
| luminous flux | lm (lumen) | cd sr |
| magnetic field strength | A/m | – |
| magnetic flux | Wb (weber) | $(kg\, m^2)/(A\, s^2)$ |
| magnetic flux density | T (tesla) | $kg/(A\, s^2)$ |
| moment of force | Nm | – |
| permeability | H/m | – |
| permittivity | F/m | – |
| power | W (watt) | $(kg\, m^2)/s^3$ |
| pressure | Pa (pascal) or $N/M^2$ | $kg/(m\, s^2)$ |
| specific volume | $m^3/kg$ | – |
| surface tension | N/m | – |
| thermal conductivity | W/(m K) | – |
| velocity | m/s | – |
| volume | $m^3$ | – |

Some units generally accepted by science and engineering circles but not formally part of the SI system are shown in the next table.

| *Physical property* | *Unit symbol* | *Unit name* | *Definition* |
|---|---|---|---|
| angle | ° | degree | (π/180) rad |
| | ′ | minute | (π/10800) rad |
| | ″ | second | (π/648000) rad |
| area | ha | hectare | 10 000 $m^2$ |
| | °F | degree Farenheit | (1.8 × °C) + 32 |
| dynamic viscosity | P | poise | $10^{-1}$ Pa s |
| energy | cal | calorie | ~4.18 J |
| force | kgf | kilogramme force | ~9.807 N |
| kinematic viscosity | St | Stokes | $10^{-4}$ $m^2/s$ |
| length | in or ″ | inch | 2.54 × $10^{-2}$ m |
| | ft or ′ | foot | 12 in |
| | μm | micron | $10^{-6}$ m |
| | Aå | angstrom | $10^{-10}$ m |
| mass | lb | pound | ~0.454 kg |
| | t | tonne | $10^3$ kg |
| pressure | atm | atmosphere | 101 325 Pa |
| | bar | bar | $10^5$ Pa |
| | mm Hg | millimetre of mercury | ~133.322 Pa |
| | torr | torr | ~133.322 Pa |
| temperature | °C | degree Celsius | K – 273.2 |
| thermodynamic temperature | °R | degree Rankine | °F + 459.7 |
| time | min | minute | 60 s |

Index notation may be used, e.g. $kg/m^3$ may be written as $kg\,m^{-3}$ and W/(m K) as $W\,m^{-1}\,K^{-1}$. There should be a space between units, e.g. A s and not As.

### *11.2.4 Unit conversion factors – imperial–metric and metric–imperial*

| *Property* | *Imperial to metric* | | *Metric to imperial* | |
|---|---|---|---|---|
| Acceleration | 1 ft/s$^2$ | 0.3048 m/s$^2$ | 1 m/s$^2$ | 3.2808 ft/s$^2$ |
| Angular velocity | 1 rev/min | 0.10472 rad/s$^2$ | 1 rad/s$^2$ | 9.5493 rev/min |
| Area | 1 in$^2$ | 6.4516 cm$^2$ | 1 cm$^2$ | 0.155 in$^2$ |
| | 1 ft$^2$ | 0.092903 m$^2$ | 1 m$^2$ | 10.764 ft$^2$ |
| | 1 yd$^2$ | 0.83613 m$^2$ | 1 m$^2$ | 1.196 yd$^2$ |
| | 1 acre | 0.40469 ha | 1 ha | 2.4711 acre |
| Density | 1 lb/ft$^3$ | 16.018 kg/m$^3$ | 1 kg/m$^3$ | 0.062428 lb/ft$^3$ |
| Energy | 1 ft pdl | 0.042140 J | 1 J | 23.730 ft pdl |
| | 1 ft lbf | 1.35582 J | 1 J | 0.73756 ft lbf |
| | 1 kW h | 3.6 MJ | 1 MJ | 0.27778 kW h |
| | 1 therm | 0.10551 GJ | 1 GJ | 9.4781 therm |
| Force | 1 pdl | 0.13826 N | 1 N | 7.233 pdl |
| | 1 lbf | 4.4482 N | 1 N | 0.22481 lbf |
| Length | 1 in | 25.4 mm | 1 cm | 0.3937 in |
| | 1 ft | 0.3048 m | 1 m | 3.2808 ft |
| | 1 yd | 0.9144 m | 1 m | 1.0936 yd |
| | 1 mi | 1.6093 km | 1 km | 0.62137 mi |
| Mass | 1 oz | 28.35 g | 1 g | 0.035274 oz |
| | 1 lb | 0.45359 kg | 1 kg | 2.2046 lb |
| | 1 cwt | 50.802 kg | 1 kg | 0.019684 cwt |
| | 1 ton | 1.0161 tonne | 1 tonne | 0.98421 ton |
| Moment of force | 1 lbf ft | 1.3558Nm | 1 Nm | 0.73756 lbf ft |
| Plane angle | 1° | 0.01745 rad | 1 rad | 57.296° |
| Power | 1 ft lbf/s | 1.3558 W | 1 W | 0.73756 ft lbf/s |
| | 1 hp | 0.7457 kW | 1 kW | 1.341 hp |
| Pressure and stress | 1 in Hg | 33.864 mbar | 1 mbar | 0.02953 in Hg |
| | 1 lbf/in$^2$ | 6.8948 kPa | 1 kPa | 0.14504 lbf/in$^2$ |
| | 1 tonf/in$^2$ | 15.444 N/mm$^2$ | 1 N/mm$^2$ | 0.064749 tonf/in$^2$ |
| Specific heat capacity | 1 Btu (ib °F) | 4.1868 kJ/(kg °C) | 1 kJ/(kg °C) | 0.23885 Btu (ib °F) |
| Velocity | 1 ft/s | 0.30480 m/s | 1 m/s | 3.2808 ft/s |
| | 1 mi/h | 1.6093 km/h | 1 km/h | 0.62137 mi/h |
| Volume | 1 in$^3$ | 16.387 cm$^3$ | 1 cm$^3$ | 0.061024 in$^3$ |
| | 1 ft$^3$ | 0.028317 m$^3$ | 1 m$^3$ | 35.315 ft$^3$ |
| | 1 yd$^3$ | 0.76456 m$^3$ | 1 m$^3$ | 1.308 yd$^3$ |
| | 1 pt | 0.56826 litre | 1 litre | 1.7598 pt |
| | 1 gal | 4.546 litre | 1 litre | 0.21997 gal |

### 11.2.5 Common scientific, mathematical and engineering constants

| *Constant* | *Value* |
|---|---|
| Avagadro's number ($N_A$) | $6.023 \times 10^{26}$ kmol$^{-1}$ |
| Boltzmann's constant ($k$) | $1.3805 \times 10^{-23}$ J/K |
| Planck's constant ($h$) | $6.626 \times 10^{-34}$ Js |
| Gravitational acceleration ($g$) | 9.807 m/s$^2$ |
| Universal gas constant ($R$) | 8.314 kJ/kmol K |
| Stefan–Boltzmann constant ($\sigma$) | $5.6697 \times 10^{-8}$ W/m$^2$ K$^4$ |
| Universal constant of gravitation ($G$) | $6.67 \times 10^{-11}$ N m$^2$/kg$^2$ |
| Electronic charge ($e$) | $1.602 \times 10^{-19}$ C |
| Electron volt (eV) | $1.602 \times 10^{-19}$ J |
| Velocity of light in a vacuum ($c$) | $2.99795 \times 10^{8}$ m/s |
| Volume of 1 kmol of ideal gas at s.t.p. | 22.41 m$^3$ |
| Standard temperature and pressure s.t.p. | 273.15 K and $1.013 \times 10^{5}$ N/m$^2$ |

| *Constant* | *Approximate value* |
|---|---|
| Density of water | ~$10^3$ kg/m$^3$ |
| Viscosity of water (at 18°C) | ~$10^{-3}$ N s/m$^2$ |
| Heat capacity of water | ~4 kJ/kg K |
| Thermal conductivity of water | ~0.6 W/m K |
| Latent heat of boiling water | ~2 MJ/kg |
| Density of air at s.t.p. | ~1 kg/m$^3$ |
| Viscosity of air at s.t.p. | ~$1.7 \times 10^{-5}$ N s/m$^2$ |
| Heat capacity of air at s.t.p. | ~1 kJ/kg K |
| Thermal conductivity of air at s.t.p. | ~0.024 W/m K |

# Index

*Index*